Conceptual Blockbusting

A GUIDE TO BETTER IDEAS

Third Edition

James L. Adams

ADDISON-WESLEY PUBLISHING COMPANY, INC.

Reading, Massachusetts • Menlo Park, California
Don Mills, Ontario • Wokingham, England • Amsterdam
Sydney • Singapore • Tokyo • Madrid • Bogotá
Santiago • San Juan

Library of Congress Cataloging-in-Publication Data

Adams, James L.
Conceptual blockbusting.

Bibliography: p.
Includes index.
1. Problem solving. 2. Creative thinking.
3. Concepts. I. Title.
BF441.A28 1986 153.4 85-28722
ISBN 0-201-10149-1
ISBN 0-201-10089-4 (pbk.)

Set in 10 point Sabon

This book was published originally as a part of THE PORTABLE STANFORD, a book series published by the Stanford Alumni Association, Stanford, California.

CONTENTS

ILLUSTRATIONS AND QUOTATIONS

FACING PAGE iii — Magritte, René. *The Thought Which Sees.* 1965. Graphite. Collection, The Museum of Modern Art, New York. Gift of Mr. and Mrs. Charles B. Benenson.

PAGE xiv — Escher, M. C. *Belvedere.* Lithograph. © M. C. Escher Heirs. Reproduced by permission of Cordon Art, Baarn, Holland. [Notice the peculiar pillars]

PAGE 2 — Arnold Palmer "Athlete of the Decade" illustration. By permission of Arnold Palmer and United Feature Syndicate.

PAGE 5 — Goldberg, Rube. *Sure Cure for Nagging Wife (or Husband) Who Continually Criticizes Your Driving; Simple Reducing Machine.* Both © 1936. Ink and watercolor. Reproduced with permission of The McNaught Syndicate, New York.

PAGE 6 — Goldberg, Rube. *Be Your Own Dentist; Professor Butts ("You Sap, Mail that Letter"); Professor Butts' Brain Takes a Nosedive.* All © 1930. Pen and ink. Reproduced with permission of The McNaught Syndicate, New York.

PAGE 8 — Drawing by Jim M'Guinness. Other drawings by Jim M'Guinness are found on pages 26, 27, 54, 86, 87, 110, and 111.

PAGE 12 — Lichtenstein, Roy. *Cathedral, #5.* 1969. Two-color lithograph. © Gemini, G.E.L., Los Angeles.

PAGE 14 — Paradoxical figures from *Eye and Brain* by R. L. Gregory, McGraw-Hill Paperbacks. © 1966 R. L. Gregory. Courtesy of the author.

PAGE 14 — "Three Legged Pants" illustration. Courtesy of Levi Strauss & Co., San Francisco.

PAGE 15 — Diagram adapted from Roberta Klatzky, *Human Memory.* Reproduced by permission of W. H. Freeman and Company.

PAGE 16	Swarm of A's. Illustration by Carol Leudesdorf.
PAGE 19	Escher, M. C. *Waterfall.* Lithograph. © M. C. Escher Heirs. Reproduced by permission of Cordon Art, Baarn, Holland. [Notice the flow of water in this perpetual motion machine.]
PAGES 21 & 22	Drawings from *Creative Behavior Workbook* by Sidney J. Parnes. Used by permission of Charles Scribner's Sons. Copyright © 1967 Charles Scribner's sons.
PAGE 21	Nicholason, Karl. Volkswagen drawing from CRM Books, *Involvement in Developmental Psychology Today.* © 1971, CRM. Reprinted by permission of Random House, Inc.
PAGE 38	Picabia, Francis. *Parade Amoureuse.* 1917. Oil on canvas. Collection Mr. and Mrs. Morton G. Neumann, Chicago. (Photo by Geoffrey Clements.)
PAGE 40	de Chirico, Giorgio. *The Anxious Journey.* 1913. Oil on canvas. Collection, The Museum of Modern Art, New York. Acquired through the Lillie P. Bliss Bequest.
PAGE 44	Unknown artist. 18th century. *Victorian Making His Flight.* Frontispiece of Restif de la Bretonne, *La Decouverte australe, par un Hommevolant* . . . Leipzig, 1781. Engraving. Courtesy of Rare Book Division, The New York Public Library, Astor, Lenox and Tilden Foundations.
PAGE 49	Hogarth, William. *False Perspectives.* Published in *The Works of William Hogarth, from the Original Plates Restored by James Heath* . . . London. 1835. Courtesy of Department of Rare Books and Special Collections, University of Michigan Library, Ann Arbor.
PAGE 52	Tycho Brahe's quadrant, c. 1576. Illustration from Tycho Brahe's *Astronomiae Instaurate Mechanicae,* 1602. Courtesy of The Science Museum Library, London. British Crown Copyright.
PAGE 56	Seymour, Robert (Shortshanks). British, 1798–1836. *Locomotion.* Cartoon, hand-colored etching. Courtesy of the Metropolitan Museum of Art. Gift of Paul Bird, Jr. 1962.
PAGE 70	Christo. *5,600 Cubic Meter Package.* Project. 1967. Pencil on paper. Collection Kimiko and John Powers, New York. (Photo and © by Harry Shunk.)
PAGE 82	Christo. *Packed Chairs and Table.* Project. 1965. Private Collection U.S. (Photo by Harry Zucker.)
PAGE 92	Drawing courtesy of Walter Thomason, San Francisco.

PAGE 93 Drawings courtesy of Peter Dreissigacker, Stanford.

PAGE 101 Unknown British artist. Mid-19th century. *The Flight of Intellect.* Lithograph, G. E. Madeley, London. Courtesy Princeton University Library, Princeton. The Harold Fowler McCormick Collection of Aeronautica.

PAGE 102 Vasarely, Victor. *Tau-Ceti.* 1955–65. Courtesy of Victor Vasarely and the Vasarely Center, N.Y.

PAGES 114 & 115 Goldberg, Rube. *Professor Butts, Training for the Olympic Games.* © 1932. Pen and ink. Reproduced with permission of The McNaught Syndicate, New York.

PAGE 121 Riley, Bridget. *Fall.* 1963. Courtesy of Bridget Riley and Tate Gallery, London.

PAGE 130 Magritte, René. *Les Mois des Vendanges.* Ink drawing. Collection of Harry Torczyner, New York. © 1985 by Georgette Magritte.

PAGE 135 Morris, Robert. *Untitled, 1967–68.* Private Collection. Photo courtesy of the Leo Castelli Gallery, New York.

PAGES 139 & 140 Charts adapted from George M. Prince, "Synectics: Twenty-Five Years of Research Into Creativity and Group Process." © 1983, The American Society for Training and Development.

PAGE 142 Chart adapted from Teresa M. Amabile, *The Social Psychology of Creativity.* © 1983, Springer-Verlag New York, Inc.

PAGE 147 Image d'Epinal. *Battle of the Pyramids,* 1798. Reproduced with permission of the Musée Carnavalet, Paris. Photographie Giraudon.

Permission to quote material from various books was obtained from the following sources:

PAGES 50 & 51 Reprinted by permission of Walker and Company from *Put Your Mother on the Ceiling* by Richard de Mille. Copyright 1955, 1957, Walker and Company, New York. [Passage on "breathing"]

PAGES 74 & 75 Reprinted by permission of David Straus from *Strategy Notebook.* Copyright 1972, Interaction Associates, Inc., San Francisco. [Strategies and page on "Eliminate"—including redrawing of figure]

PAGES 105, 109 & 110 Reprinted by permission of William Kaufmann, Inc., from *The Universal Traveler* by Don Koberg and Jim Bagnall. © 1974 by William Kaufmann, Inc. All rights reserved. [Passage on "Constructive Discontent" and passage on "Morphological Forced Connections"—including redrawing of "cube pen"]

PAGES 114 & 115 Reprinted by permission of Charles Scribner's Sons from *Applied Imagination* by Alex F. Osborn. Copyright © 1963, Charles Scribner's Sons, New York, 3rd revised edition. [Checklist for New Ideas]

PAGES 116 & 117 Reprinted by permission of Princeton University Press from *How to Solve It: A New Aspect of Mathematical Method,* 2nd ed., by G. Polya. Copyright 1945, Princeton University Press, © 1957 by G. Polya. A selection on pp. xvi, xvii.

PAGES 120, 122 & 123 Reprinted by permission of Harper & Row, Publishers, Inc., from pages 49–51 in *Synectics* by William J. J. Gordon. Copyright © 1961 by William J. J. Gordon, New York. [A selection on Fantasy Analogy]

PAGE 138 Reprinted by permission of Jonathan Prince. © 1980, 1985 by Synectics, Inc. [Sample Synectics "excursion"]

PAGES 144 & 145 Reprinted by permission of Springer-Verlag New York, Inc., from *The Social Psychology of Creativity* by Teresa M. Amabile. © 1983 by Springer-Verlag New York, Inc. [Passage on creativity of collage artists given "specific creativity instructions"]

Magritte

PREFACE

Few people like problems. Hence the natural tendency in problem-solving is to pick the first solution that comes to mind and run with it. The disadvantage of this approach is that you may run either off a cliff or into a worse problem than you started with. A better strategy in solving problems is to select the most attractive path from many ideas, or concepts. This book is concerned with the cultivation of idea-having and problem-solving abilities.

Since I am a teacher of and consultant in engineering and management, the majority of my contacts are with people who are analytical, quantitative, verbal, and logical. These are certainly excellent traits for any problem-solver. However, so is the ability to conceptualize freely, an activity that requires a somewhat broader thinking vocabulary. I am deeply concerned with attempting to better define this vocabulary and with helping people expand in this direction. This book reflects that concern. Although I originally began this quest in an engineering environment, my experience with people in other fields has convinced me that the material in this book is applicable to most walks of life.

You will find little within about how to be more verbal or analytical, although these are both powerful problem-solving tools. Instead the book concentrates upon aspects of thinking that are useful in improving one's conceptual abilities and that I feel are underemphasized in many people's education. Although the book focuses upon conceptualization, these aspects of thinking are also pertinent in other parts of the problem-solving process. Much of the book's emphasis is on creativity, since a good conceptualizer must be a creative conceptualizer. The mental characteristics that seem to make one creative not only are valuable in idea-having, but also better equip one to find and define problems and implement the resulting solutions.

The material herein draws upon a multitude of sources. I was first introduced to thinking about thinking by the late John E. Arnold, a pioneer in education and one of my all-time personal heroes. Quite a bit of the book reflects his thinking. Another major influence was Professor Bob McKim, a colleague and friend. We logically should have written this book together. However, Bob had just completed an excellent book of his own and was temporarily down with an overexposure to writing. Nonetheless, his influence and thoughts appear throughout. Bob also had the stamina to read the manuscript, as did Professor Harold Leavitt of the Stanford Graduate School of Business, an authority on organizational behavior, and Dr. James Fadiman, lecturer and resident psychologist of the Design Division. Cynthia Fry Gunn, my editor, transformed the manuscript into readable prose and selected the art work which adorns the following pages. Special thanks are due to Brooks/Cole Publishing Company for permitting the use of material contained in McKim's *Experiences in Visual Thinking.*

Thinking is not yet fully understood. You will therefore find material in this book based on unproven theory, conjecture, and non-scientific observation. Feel free to disagree with it if you would like to, because the intent of the book is not to deliver to you the last word in psychological theory. Its aim is rather to let you learn something about how your own mind works in a conceptual situation and to give you some hints on how to make it work better. This is, accordingly, a "think along" book, since it is difficult to talk about thinking without observing one's own thoughts. It is also a book that will, I hope, stimulate you to dig deeper into the game of thinking, with resulting benefits to your own idea-having capability.

James L. Adams

Stanford, California
March 5, 1974

NOTE TO THE SECOND EDITION

A chapter has been added on groups and organizations, additional material has been added on perception, and the content has been somewhat updated. However, the basic message of the book and much of the original material remain unchanged.

J. L. A.

July 11, 1979

NOTE TO THE THIRD EDITION

It's been over ten years since *Conceptual Blockbusting* was first published. I am even more convinced of the value of consciously identifying conceptual blocks than I was when I originally wrote the book. The process is not only interesting in its own right, but is a powerful tool in increasing creativity. As we enter the latter half of the nineteen-eighties, it is becoming increasingly clear that new ways of thinking and of engendering creativity need to become a natural part of the way we live.

We are quite programmed in our thinking for a number of good reasons, and tend to have well-developed thinking styles. Therefore we all have conceptual blocks. However, we are fortunately blessed with the ability to consciously modify our problem-solving habits in order to arrive at more creative inputs. In order to do this we must acquire increased awareness of our problem-solving process. We need to become suspicious of our business-as-usual habits and learn to better recognize when we should modify them. We need to learn more about the details of creative problem-solving. This book, with its emphasis on mental blocks, is an effective way toward such self-knowledge since conceptual blocks are universal, easily identifiable, and can, with effort, be modified.

Everyone wants to be more creative (or thinks they do). Motivation is not a problem. An understanding of conceptual blocks can only increase your motivation, simply because these blocks are not consistent with your self image (Who, me? Stereotype? No way!). In addition to this internal motivation, life seems to give us many additional reasons to be creative. The combination of motivation and conscious intervention based on increased awareness and knowledge of the problem-solving process is the standard formula for increasing creativity. This book seems to be successful enough in helping people accomplish this that I am not about to make major changes in its message.

For this third edition, I have updated the material and expanded the final chapter. In particular, I have included more discussion of the relative roles of internal and external motivation in creativity, since I'm convinced that the only way to become a consistently better problem-solver is to understand both the how *and* why of creativity.

October 15, 1985 J.L.A.

CHAPTER ONE

INTRODUCTION

OUR LEARNING CAPACITY is truly impressive, not only in terms of knowledge, but also in terms of function. The amount of information the brain can retain is phenomenal. However, so is its ability to control the actions of the juggler, the stunt pilot, or the musician. Some of our functions, such as circulating blood and sensing temperature, although magnificent in complexity, are automatic in that they do not require conscious learning. Others, such as running and vocalizing, are easily, almost naturally acquired, yet demand considerable conscious effort if a level of excellence is to be attained. Still others, such as tennis, leather tanning, espaliering, chess playing, hang-gliding, and reading, must be acquired through conscious effort.

What about thinking? It is certainly a most important function. Is it automatic? Is it learned consciously? The time-honored method of improving one's skill is to be continually conscious of one's performance and to seek to improve it—usually according to an ideal or standard of what is desirable. The serious golf player studies golf and then continually practices, comparing his performance and form against an ideal, reading books and newspaper columns on golf form, and watching other more sophisticated golfers as they play.

Should the thinker act like this? Should we learn all we can about thinking and then practice and monitor our result? Should we compare our thinking with that done by more sophisticated thinkers? The common piece of Americana pictured below is part of a series designed to help golfers improve their performance.

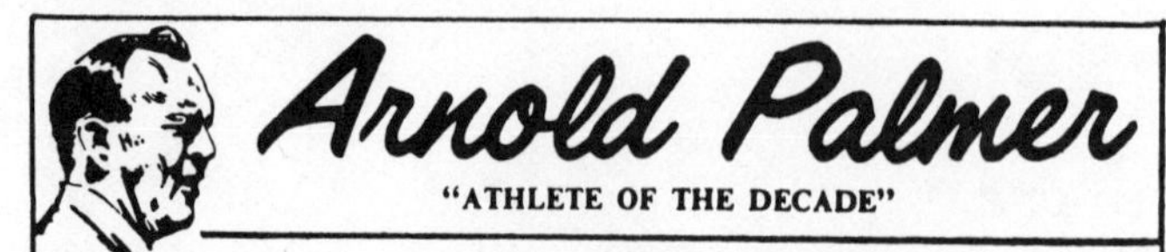

ROLL INSIDE ON LEFT FOOT

The manner in which you lift your left foot during your backswing influences the success of your over-all swing.

If you go up on the toe of this foot, as the golfer in illustration No. 1 is doing, you react in a reverse manner on

your downswing. You will lower the heel in a manner that shifts weight to the back of this foot. This could cause you to fall back on your heels.

If you roll onto the inside of your left foot during your backswing, however (illustration No. 2), you will tend to react on your downswing by shifting your weight to the left (illustration No. 3). This is the proper weight transfer that is so necessary for consistent shot-making.

Have you ever seen a similar treatment of thinking? All of us are thinkers. However, most of us are surprisingly unconscious of the process of our own thinking. When we speak of *improving the mind* we are usually referring to the acquisition of information or knowledge, or to the *type* of thoughts one *should* have, and not to the actual *functioning* of the mind. We spend little time monitoring our own thinking and comparing it with a more sophisticated ideal.

Thinking Form

There are reasons for this, of course. Thinking "form" is much more difficult to observe than, say, golfing "form." Thinking is also a much more complex function than golf. If you were to write the thinking analogue to the golf column, how could you select the "thinker of the decade" and how could you extract as simple an element as the roll of the left foot from the complex process of thinking? Yet, despite these problems, effort spent in monitoring the thinking process and attempting to improve it is a good investment for the problem-solver.

Much of thinking must and does occur in a spontaneous (unconscious) way, with little energy consciously expended in monitoring or seeking to improve the process. Let me give you an example which may illustrate this. However, before I do, let me diverge a moment and make a general comment about this book. It contains occasional examples and exercises. The content of the book is much more meaningful and much more likely to influence your thinking if the exercises and problems are worked. You can do this either alone or with other people. I have found that most of them are usually more entertaining and more successful if several people are involved. It is always of interest to see the variation in thinking among a number of people. Try the exercises on your friends and associates at whatever occasion may seem appropriate, whether they are reading the book or not. In any case, try to work them yourself. You will need only paper and pencil. It is surprisingly easy to read material about thinking, accept it intellectually, and yet not have one's own thinking processes affected. This book is a little bit like one about jogging. It won't do you nearly as much good unless you run a little.

Now, back to the example. The following puzzle, which originates with Carl Duncker, is taken from *The Act of Creation* by Arthur Koestler. Work on it awhile. When you get the answer or get tired of thinking about it, proceed.

Puzzle: "One morning, exactly at sunrise, a Buddhist monk began to climb a tall mountain. A narrow path, no more than a foot or two wide, spiraled around the mountain to a glittering temple at the summit. The monk ascended at varying rates of speed, stopping many times along the way to rest and eat dried fruit he carried with him. He reached the temple shortly before sunset. After several days of fasting and meditation he began his journey back along the same path, starting at sunrise and again walking at variable speeds with many pauses along the way. His average speed descending was, of course, greater than his average climbing speed. Prove that there is *a spot* along the path that the monk will occupy on both trips at precisely the same time of day."

Did you solve the puzzle? More importantly, for our purposes, can you remember *what thinking processes* you used in working on the puzzle? Did you verbalize? Did you use imagery? Mathematics? Did you consciously try different strategies or attacks on the problem? It is probable that you tried several methods of working the problem, but that your mind automatically switched from one to the other. You were probably not particularly aware of what mental processes you were employing as you thought about the problem. You were playing a game (like tennis) without being very aware of what you were doing or of techniques by which you could improve your game (like getting your racket back faster).

A simple way of solving the puzzle is to visualize the upward journey of the monk superimposed upon the downward journey. Visualize, if you would, two monks, one at the bottom of the path and one at the top as the sun is rising. Let the bottom monk duplicate the upward journey as the upper monk duplicates the downward journey. It should be apparent that at some time and at some point on the path they will collide. This point is the spot along the path and the time of the collision is the time.

If you happened to choose visual imagery as the method of thinking to apply to this problem, you probably solved it. (A slightly more abstract approach is to imagine a plot on a graph of each monk's position as a function of time. The two lines will necessarily cross at a common position and time.) If you chose verbalization, you probably did not solve the problem. In fact, even after knowing the visual solution, if you revert to a verbal attack, the problem becomes confusing again. If you attempted an abstract mathematical approach that did not involve graphing, you probably once again failed to solve the problem and expended much more effort than was necessary.

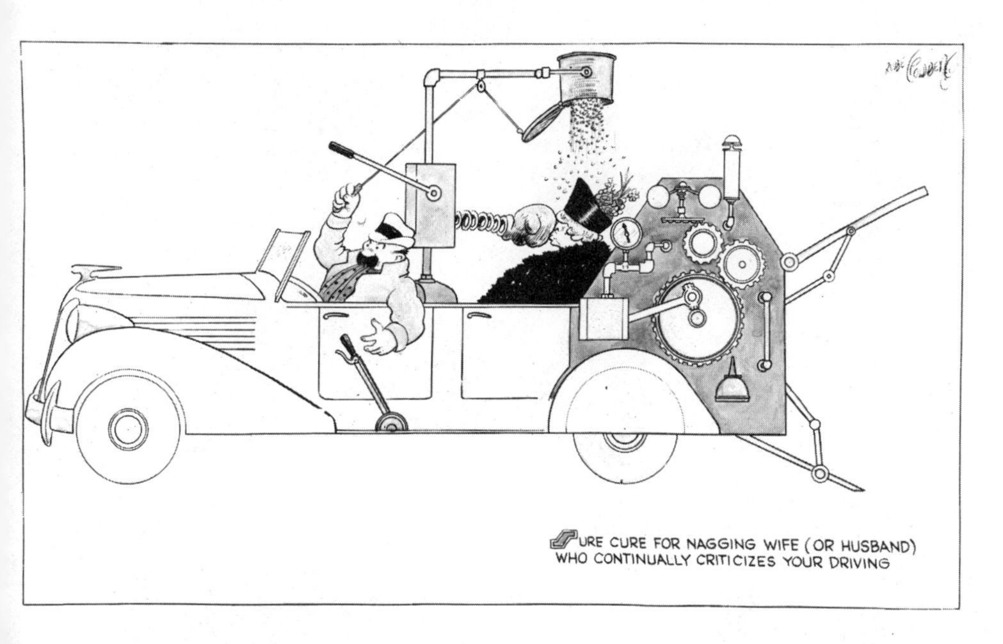

BE YOUR OWN DENTIST!

FIRST TIE YOURSELF SECURELY TO CHAIR (A) AND WIGGLE FOOT (B). FEATHER (C) TICKLES BIRD (D) - AS BIRD SHAKES WITH LAUGHTER, IT MIXES COCKTAIL IN SHAKER (E) - BIRD FALLS FORWARD, SPILLING COCKTAIL, AND SQUIRREL (F) GETS SOUSED - IN HIS DRUNKEN EXCITEMENT, SQUIRREL REVOLVES CAGE (G), WHICH TURNS CRANK (H) AND PLAYS PHONOGRAPH RECORD (I) - SONG (J) GETS DWARF (K) HOT UNDER COLLAR AND FLAMES (L) IGNITE FUSE (M) WHICH SETS OFF CANNON (N), SHOOTING OUT CANNON BALL (O), CAUSING STRING (P) TO PULL TOOTH!

PROFESSOR BUTTS GETS CAUGHT IN A REVOLVING DOOR AND BECOMES DIZZY ENOUGH TO DOPE OUT AN IDEA TO KEEP YOU FROM FORGETTING TO MAIL YOUR WIFE'S LETTER.

AS YOU WALK PAST COBBLER SHOP, HOOK (A) STRIKES SUSPENDED BOOT (B) CAUSING IT TO KICK FOOTBALL (C) THROUGH GOAL POSTS (D). FOOTBALL DROPS INTO BASKET (E) AND STRI (F) TILTS SPRINKLING CAN (G) CAUSING WATER TO SOAK COAT TAILS (H). AS COAT SHRINKS CORD (I) OPENS DOOR (J) OF CAGE ALLOWING BIRD (K) TO WALK OUT ON PERCH (L) AND GRAB WORM (M) WHICH IS ATTACHED TO STRING (N). THIS PULLS DOWN WINDOW SHADE (O) ON WHICH IS WRITTEN, **"YOU SAP, MAIL THAT LETTER."** A SIMPLE WAY TO AVOID ALL THIS TROUBLE IS TO MARR A WIFE WHO CAN'T WRITE.

PROFESSOR BUTTS' BRAIN TAKES A NOSEDIVE AND OUT COMES HIS SELF-WATERING PALM TREE.

STRING (A) WORKS JUMPING JACK (B), FRIGHTENING CAT (C) WHICH RAISES BACK AND LIFTS TROUGH (D), CAUSING BALL (E) TO FALL INTO TEACUP (F). SPRING (G) MAKES BALL REBOUND INTO CUP (H) PULLING ON STRING (I) WHICH RELEASES STICK (J), CAUSING SHELF (K) TO COLLAPSE. MILK CAN (L) DROPS ON LADLE (M) AND TENSION ON STRING (N) TILTS SHOE (O) AGAINST JIGGER ON SELTZER BOTTLE (P), SQUIRTING SELTZER ON ASH-CAN SPANIEL WHO HASN'T HAD A BATH IN FOUR YEARS. SURPRISE CAUSES HIM TO TURN THREE SOMERSAULTS OVER APPARATUS (R) AND WATER SPLASHES NATURALLY INTO BOWLS, RUNNING THROUGH SPRAY (S), WATERING PALM (T), AND SAVING YOURSELF A TRIP TO HAVANA FOR TROPICAL ATMOSPHERE.

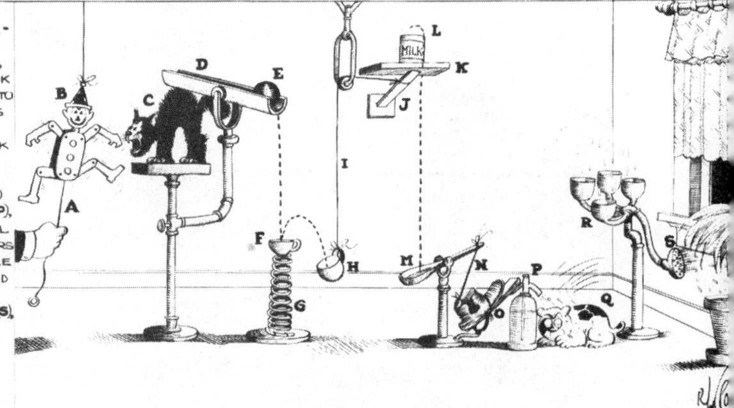

This book aims to make you more aware of what is going on in your mind (what you are doing) and to give you a few techniques (getting your racket back faster) which may improve your capability to solve problems (the game). We will be concentrating upon *conceptualization*, or the process by which one has ideas. This process is a key one in problem-solving, since the more creative concepts you have to choose from, the better. This is true at all stages of the problem-solving process, whether you are attempting to decide upon a broad direction or implement a detailed solution.

By concentrating on conceptualization, I am not attempting to downgrade the many other processes necessary in problem-solving, such as judgment, analysis, proper problem-definition, and the critical aspect of coaxing the idea into reality. Neither am I trying to insult your intelligence by pointing out the obvious value of a rich store of concepts to choose from. However, my work with students, professional people, and others over the years has convinced me that conceptualization does not always receive the attention it should in problem-solving. Conceptualization in problem-solving should be creative and should be treated as a major activity. Unfortunately, in actual problem-solving situations, people often fall short of this goal.

As mentioned earlier, the natural response to a problem seems to be to try to get rid of it by finding an answer—often taking the first answer that occurs and pursuing it because of one's reluctance to spend the time and mental effort needed to conjure up a richer storehouse of alternatives from which to choose. This hit-and-run approach to problem-solving begets all sorts of oddities—and often a chain of solution-causing-problem-requiring-solution, *ad infinitum*. In engineering one finds the "Rube Goldberg" solution, in which the problem is solved by an inelegant and complicated collection of partial solutions. I am sure that many of you are familiar with some example of this in the form of an appliance you have attempted to repair.

Solutions To Problems That Don't Exist

In problem-solving, we also encounter solutions to problems that do not really exist: the objects represented in the drawing (page 8) illustrate this situation. They are devices that retard the opening of solar panels for spacecraft. The two on the right were developed by an extremely competent group (of which I was a part) in conjunction with the development of the Mariner IV, which was the first spacecraft to fly by Mars. The Mariner IV was to be provided with electrical power by four solar panels, which were to be latched together during launch, and then released and opened by spring-loaded actuators. Since there is no air in space to damp the opening of such panels and since they were covered with fragile and expensive solar cells, it was the custom to use a device to retard their opening.

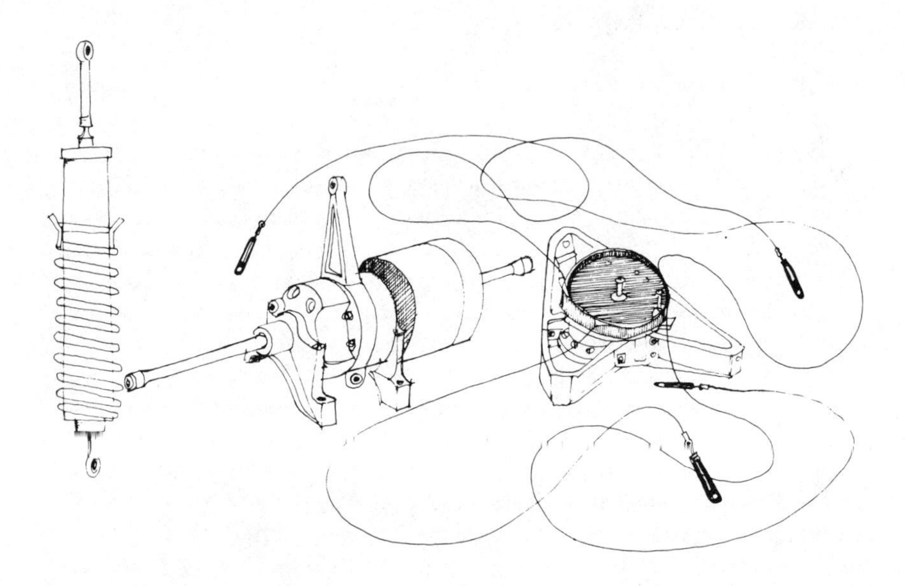

The object on the left of the figure is such a device, which was used successfully on earlier lunar spacecraft. However, it was heavy and the designers of the Mars spacecraft did not trust it since it was filled with oil and had the potential of coating the spacecraft with a lethal layer of slime during the nine-month journey to Mars. The object in the middle of the drawing was the first solution to the problem. Unfortunately, although it contained no oil, it was extremely complex and was no lighter than the previous retarders. Its complexity and the results from a large amount of testing resulted in its rejection on the grounds of inadequate reliability.

The object on the right was the second solution. This was a central retarder that would control the opening speed of all four panels. Although it was filled with oil, the oil could not leak and the device was light in weight. However, it also proved to be unreliable as originally developed. At this point, full panic occurred in the program. There was no longer time to try a third approach, since planetary spacecraft cannot be delayed (the planet becomes much more difficult to reach until the next favorable alignment of the solar system, which usually does not occur for several years). An extremely expensive, around-the-clock emergency effort was therefore launched to increase the reliability of the damper, along with a simultaneous program of testing in order to measure the adverse effects of various malfunctions of the central retarder which might occur in flight.

One of the malfunctions investigated was that in which the retarder failed completely to retard. Amazingly enough, the results were acceptable. The retarders were not, in fact, necessary at all. It was possible to allow the panels to open free and to catch them with energy absorbers at the end of their travel. The final solution to the problem is therefore illustrated by the very *absence* of a fourth object in the figure, since Mariner IV went to Mars without retarders—the most elegant possible solution to the problem.

The moral in the story is obvious. The apparent shortage of time in the development of this project coupled with the natural desire of those involved to solve problems as quickly as possible resulted in overlooking alternative concepts (such as *no* retardation) that could have prevented the wild-goose chase.

Examples of the effects of the hit-and-run approach are as plentiful in other fields. Perhaps the most dramatic are hastily thought-out and implemented solutions that create more problems than they solve. Many examples of this have been given publicity in the environmental area and have resulted in the layers of regulation that now apply to anyone working on problems with environmental impact. Some of this regulation will undoubtedly be beneficial, since it requires that a large amount of effort be put into conceptualization before a solution is implemented. The amount of thought that has gone into the problem of moving oil from the north slope of Alaska to the rest of the U.S. would probably not have occurred without governmental pressure. The quality of the solution has benefited from this conceptualization.

The ability to conceptualize productively and creatively is as important in painting the bathroom as in moving oil from Alaska, in taking family vacations as in designing spacecraft, and in spending a family income as in protecting the environment. I am convinced that the conceptual process is a general one and that the same problems arise in thinking up a more nutritional diet as in thinking up a better way to image the heart ultrasonically.

Thinking well (once again, like playing tennis well) requires that many decisions be made unconsciously. One can no more think well by consciously picking each strategy and writing each sentence out longhand in the mind than one can play tennis well by consciously thinking of what position to place each joint in the body as one attempts to reach a difficult shot. However, just as tennis benefits from your becoming so familiar with various strategies that they become automatic, so does thinking.

As the book proceeds, I will attempt to make you more conscious of the creative process, various blocks that inhibit it, and various tricks that can augment it. Although this is not a psychology book, some of the theory underlying creativity will be briefly explored. These theoretical explications, although interesting, are incomplete. Hence, there is no specific thinking pattern that can make everyone into a superconceptualizer. Some techniques and approaches work effectively for some people and yet fill others with loathing. If you come up against techniques or exercises that do nothing for you, plunge on. You will (I hope) soon find techniques and exercises that prove interesting to you.

A question that always arises when one approaches the *teaching* of creativity is "Can it be taught?" Obviously I think that it can be, or I would not be writing this book. The teaching may be more of an *encouraging*, but call it what you will, I am convinced that our efforts at Stanford result in an improvement in the quality of conceptual output from our students. Another question that often comes up is "Can't conscious efforts to be creative interfere with the creative process?" This will be dealt with more thoroughly in Chapter Three but, in brief, the answer is simple. If a da Vinci happens to be reading this book he is probably wasting his time. However, most of us are not da Vincis.

One of the earliest theories about creativity considers it to be a divine spark. Plato, in III, Ion, says about poets:

> And for this reason God takes away the minds of these men and uses them as his ministers, just as he does soothsayers and goodly seers in order that we who hear them may know that it is not they who utter these words of great price when they are out of their wits, but that it is God himself who speaks and addresses us through them.

However, for most of us, creativity is more of a dull glow than a divine spark. And the more fanning it receives, the brighter it will burn.

Conceptual Blocks

Let me now make a few comments about the framework of this book. Chapters Two through Five will be concerned with conceptual blocks:

mental walls that block the problem-solver from correctly perceiving a problem or conceiving its solution. Everyone has them. However, they vary in quantity and in intensity from individual to individual. Most of us are not aware of the extent of our conceptual blocks. Awareness can not only allow us to better know our strengths and weaknesses, but can give us the motivation and the knowledge necessary to modify or avoid such blocks. These four chapters will discuss conceptual blocks, giving examples and exploring their causes. The blocks are closely related, as you will see when you begin to consider them. The particular scheme used to categorize them is for convenience only, and is not meant to be the ultimate morphology of conceptual blocks. Once again, please do the exercises and problems. The only way you will identify your own conceptual blocks is to try activities that are impeded by their existence.

Chapters Six and Seven are concerned with *techniques* that allow you to overcome (or sidestep) these blocks. Chapter Eight deals with conceptualization in a group or organizational setting. The final section of the book, the Reader's Guide, contains information for those interested in pursuing this subject in greater depth. A great deal of material exists on creativity and conceptualization, and most of it is accessible without a specialist's vocabulary. I recommend it as not only fascinating, but also unique in that you are simultaneously able to acquire knowledge and improve your idea-having and problem-solving capability.

CHAPTER TWO

PERCEPTUAL BLOCKS

PERCEPTUAL BLOCKS ARE OBSTACLES that prevent the problem-solver from clearly perceiving either the problem itself or the information needed to solve the problem. Perhaps the best way of helping you overcome perceptual blocks is to talk about some common and specific ones.

One: Seeing What You Expect to See—Stereotyping

Over the past twenty years we have been continually reminded of the existence of stereotyping. Members of ethnic minorities, women, homosexuals, the elderly, the handicapped, and others have successfully taught us that social stereotypes are wrong. Yet stereotypes continue to play a large role in our lives. If you are male, imagine the effect on people's perception of you if you were to wear high-heeled shoes. If you are a woman, just light up a cigar the next time you go into a job interview. Stereotyping and labeling are extremely prevalent and effective perceptual blocks. The simple truth of the matter is, you cannot see clearly if you are controlled by preconceptions.

As another example of the power of stereotyping, I wear neckties. I do not like them and at one time considered never wearing them again. However, I decided that this was a foolish battle, because the stereotyping associated with a necktie is so strong that I can accomplish things professionally much easier by wearing one, since people assume I am more important.

Perceptual stereotyping is part of the explanation for the success of various types of optical trickery such as in the figures below:

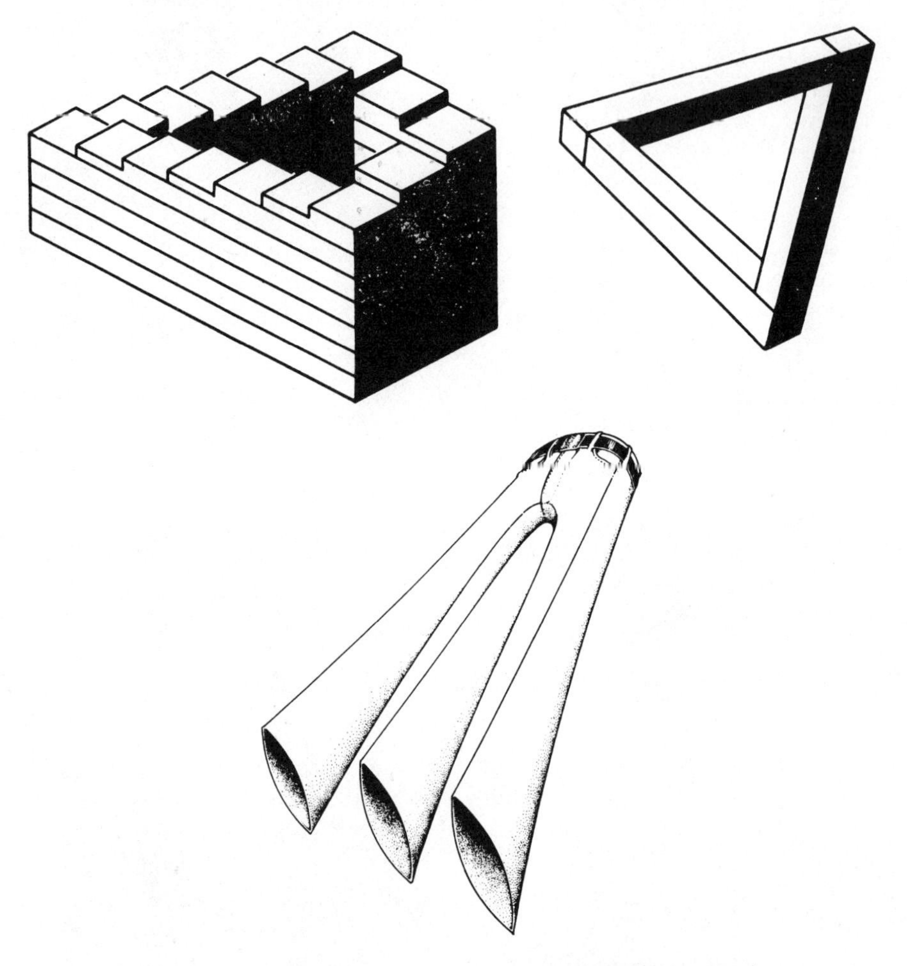

Perceptual stereotyping is not all bad, since it also allows people to complete incomplete data. However, it can be a serious handicap to perceiving new combinations. Creativity has sometimes been called the combining of seemingly disparate parts into a functioning and useful whole. Stereotyped conceptions of the parts hinder their combination into a new whole, where the roles they play may be quite different.

Once a label (professor, housewife, black, chair, butterfly, automobile, laxative) has been applied, people are less likely to notice the actual qualities or attributes of what is being labeled. For instance, say I am trying to think of what to do with a warehouseful of chairs. If I can think of them only as chairs, I can probably only come up with such uses as sitting on them, standing on them, or hitting villains with them in grade-B movies. But if I think of the attributes of the chairs (fabric, padding, wooden legs, screws, and so on), I can come up with many more uses. Maybe I should take the chairs apart and sell the seats to people who attend football games, make purses from the leather back covering, sell the screws as surplus hardware, and sell the wood to home craftsmen. Stereotyping inhibits this type of thinking.

Unfortunately, stereotyping is inherent in the working of the mind. Much of the information used in conceptualizing is first recorded in the memory and later recalled, rather than used immediately upon acquisition. The memory cannot retain all of the raw information that comes in through the senses. The mind therefore processes it by filtering out what is judged to be less useful and categorizing the rest to be as consistent as possible with information already stored in the memory. When the information is later recalled, it is in a simplified and regularized form—in a sense, a stereotype of the original.

The figure below is a cognitive psychologist's model of the information-processing system of the mind:

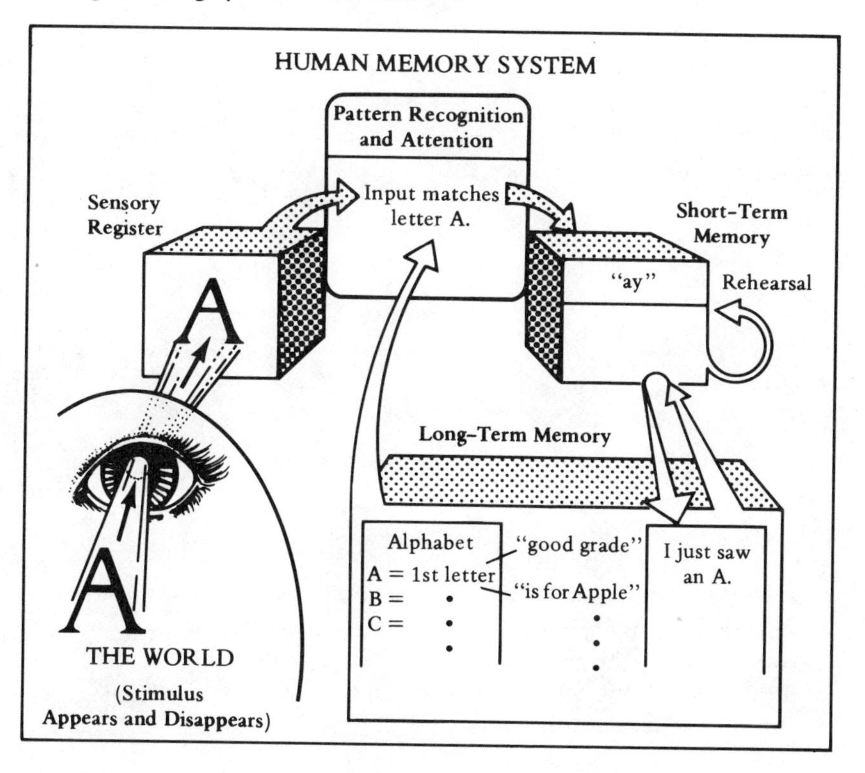

Models such as this do not correspond physically to the mind, nor is there even agreement among psychologists that the functions shown separately operate separately. However, models are a valuable aid to understanding. Let us discuss briefly the various components of this system in order to gain a better understanding of the process that causes stereotyping.

The sensory register shown in the model holds incoming information for a fraction of a second in the form received from the body's sensors. During this time, a pattern-recognition process takes place that reduces the information to a more conceptual form. For example, the pattern-recognition function in your mind allows you to discern the letter "a" from the myriad of forms that appear below (this is obviously a sophisticated process):

After the information is reduced to a more conceptual form, it goes into the short-term memory (STM). STM is much more limited than the sensory register in the amount of information it is able to store, but it can hold this information for a greater length of time (several seconds).

Although STM is extremely important in carrying out our daily activities, we are particularly interested in long-term memory (LTM), which allows us to solve problems, gives us our sense of self, and enables us to communicate sensibly. Only a small portion of the information that enters our sensory register and filters through STM ends up in LTM. *Attention* provides the focusing mechanism of LTM. While performing the complex tasks of living (such as driving to work in the morning), LTM is attending to only a small fraction of the inputs from the senses. Most of these inputs merely cycle through STM. This process of attention is shown in the above diagram by the arrow from the LTM back to the pattern-recognition function.

It is the material already in LTM that determines attention; the mind tends to reinforce what is already there. For instance, if you are an oenophile, you will record a great deal of new material you encounter on wine to add to your already considerable store of knowledge. Similarly, you will record very little information on a topic you dislike. If you hate math, you will record little new math-related information. This tendency should make you suspicious as to whether material you recall from your memory contains an honest representation of detail that is unpleasant to you. (It does not.) It should also make you suspect that stereotyping is particularly strong in areas that have been unimportant to you and/or unpleasant for you to think about. (It is.)

Information reaching LTM must be filed, and this process depends on context. The following exercise illustrates this:

> **Exercise:** Ask a friend to read aloud each set of word-pairs below, allowing about five seconds between pairs and twenty seconds between sets. Try to remember the first set of seven pairs by rehearsal (say them repeatedly to yourself). Try to remember the second set by devising a sentence that links each pair. For the third set, create a visual image associating the two words (the more bizarre the better) in your mind. After you are finished, ask your friend to read only the first word of each pair, again allowing about five seconds between words. In each instance try to write down the second word of the pair. Then check to see how many you recalled.

chair–cloud	radio–hand	snake–fireplug
fork–coin	shoe–river	chimney–boat
rug–brick	house–bug	drum–rabbit
chisel–milk	knife–flower	king–garage
church–egg	salt–bean	fish–wheel
girl–book	tire–candle	octopus–airplane
milk–star	sofa–car	cow–flower

Those words you recalled were successfully transferred to LTM. This test is not a simple one, since many people accompany their sentences with visual images and their images with subvocal sentences. However, people generally have better luck with the second set than with the first and find the third set easiest of all. The increasingly rich contexts of sentence and visual image aid the transfer of the incoming information to LTM.

Context is a key element in many memory techniques. One of the best known of these is the "method of loci." In this technique you first take a familiar walk and remember a number of scenes from the walk. To remember a number of items, superimpose a visual image of an item on each of the walk scenes. Recall then requires only mentally retracing the walk. Try it. This technique is surprisingly effective, especially for people with good visual-imagery ability. It is rumored that Cicero used this method to remember his orations to the Roman senate. It is further rumored that the technique is the origin of the phrase "in the first place we find . . . , in the second place we find . . . , etc." Probably not true, but it is such a good rumor that it should be.

We usually remember information in context, and the context goes into our memory along with the information. When we later recall the information for use in problem-solving, the residual information and feelings from the original context tend to accompany it. This complicates the conceptual process, since the residual material must be dealt with. If your first introduction to organ music is at a funeral, it may be difficult later to think of using organ music in a joyful pageant. Organ music has been, in a sense, stereotyped.

Information is also filed in the memory in a structured way. It is arranged in categories according to likely associations. Think of the word "menu." What else comes to mind? Waiters? Candles? Wine? Napkins? Lots of forks? Other restaurant scenes? Did you think of a snake? What about a tractor? A snake or tractor is unlikely, since they are not in your "restaurant" file. The structured information in your memory is so important to you that you may dismiss information that is inconsistent with that which is already there. Psychologists write about an unpleasant internal state, called cognitive dissonance, that results from an inconsistency among a person's knowledge, feelings, and behavior. The individual attempts to minimize this dissonance. One way to do this is to devalue information that does not fit one's stereotype.

With this in mind, let's look a bit further at labeling as applied to people. We all have stereotypes about people, and these often lead to social and interpersonal problems. I am a professor. Most of you, having never met me, can conclude quite a bit about me from the label "professor"

and your stereotyping ability. However, although some of the characteristics you attribute to me might be accurate, you would have trouble working or living with me with only that information, for I have my own particular group of characteristics. I am a father, an acceptable mechanic, a machinist and carpenter, a poor but enthusiastic tennis player, a better basketball player, and a pianist. I tend to be happier in rural environments than in cities, like a great deal of contact with other people in my work, but prefer a light and relaxed social schedule. I have a bad knee, a messy office, gray hair, and a 1909 brown shingle house. I am 6′4″, have green eyes, and like old machinery. As I list these attributes, you should be able to move beyond your ideas of the stereotype to get a better feel for me as a person, and therefore be better able to interact with me. I have also enriched the stereotype you have of "professor" by adding information. Now you try it. In the following exercise, see how you label yourself and how people label you.

> **Exercise:** Find someone you do not know too well. Each of you think of, and tell the other, a label (a few words) that describes you. Spend half a minute or so considering what the other person's label means to you. Then spend five minutes verbally exchanging additional characteristics. Alternate and keep moving. Do not succumb to the temptation to small talk (thereby being witty and engaging) to avoid trading information. Do not question the other person and do not try to steer the conversation. Just swap information.

Did you find this exercise a quick way to find out information about another person? Many people do. However, did you also find it difficult? Even after having lived a reasonably long and rich life, people generally run short on characteristics after a few minutes and spend more effort in thinking up their own attributes than in listening to those of the other person. They also generally experience an overwhelming desire to small-talk. Did you?

In social and professional interactions we tend to stick to stereotypes and generalities, unless at some point it seems to our benefit to become specific about ourselves. The above exercise therefore invades our privacy, since it forces us to divulge information before we may be ready to do so. After this exercise is over, most participants agree that they know much more about the other person than they would have gleaned from the original label. They also gain a sense of the importance of their own stereotype to them, as well as a better feeling of how they cling to stereotypes to avoid taking social risk. As a final comment, the exercise also shows that we do not have a large store of characteristics about ourselves in our memory. If we did, the exercise would be much easier. We not only stereotype other people and things, but we stereotype ourselves. Stereotyping is an obvious perceptual block—perhaps the most important one. But there are others, and let's now discuss some of them.

Two: Difficulty in Isolating the Problem

Can you tell what the above illustration is? If you have seen this before you probably have no trouble in discerning it. If you have not, try to identify the contents before you continue. (Answer is on next page.)

Another puzzle such as this is the picture below. What is it?

It's a Volkswagen, of course.

Now that you have seen the answer, it should be easy to see the cow. This is typical of visual puzzles that require the solver to detect meaning in the midst of apparent chaos.

Problems we face may be similarly obscured by either inadequate clues or misleading information. And proper problem-identification is of extreme importance in problem-solving. If the problem is not properly isolated, it will not be properly solved. Successful medical diagnosis depends on the ability to isolate the problem within the complexity of all of the real and imaginary information available to the physician. Successful coexistence between parents and teenage children requires the ability to isolate the real problems among many of the apparent ones.

Is your problem really a bad tank of gas, or does your car need timing or perhaps new distributor points? Or is your problem a living situation that makes you overly dependent upon a car? Problem statements are often liberally laced with answers. The answers may be well thought out or poorly conceived. They may be right or wrong. A problem statement to an architect such as "put a latch on that door between the kitchen and the dining room so that the door can be opened extremely easily" implies that the answer to kitchen/dining room access is a door, rather than no door, a redefinition of space, or a redefinition of the food preparation/eating function.

If you are working as a professional problem-solver, you must continually be alert to properly perceive the problem. The client, patient, customer, etc. may not always see his problem clearly, and the problem-solver is sometimes able to score heavily by curing the difficulties in a simpler manner through a clearer perception of what the problem really is. In engineering, people occasionally become so involved in attempting to optimize a particular device that they lose sight of alternate ways to alleviate the difficulty. Much thinking went into the mechanical design of various types of prototype tomato pickers before someone decided that the real problem was not in optimizing these designs but rather in the susceptibility of the tomatoes to damage during picking. The answer to the problem was a new plant, with tougher-skinned, more accessible fruit.

Problems are, of course, often constrained by considerations other than mere removal of a difficulty, and the problem-solver must be sensitive to this. Assume, for instance, that I am a consulting engineer retained to help in the design of an improved product by a company that was a leading manufacturer of mechanical equipment used to clear clogged drains and sewers. Assume further that I perceive the problem to be a general one of unclogging drains and sewers. This might lead me to a very elegant solution (a mixture of commonly available chemicals) that would make obsolete the product line of the company and that would not take advantage of the company's field of competence. Although I could then proudly take my place among successful conceptualizers, my employer would probably not enjoy paying me. Properly isolating the problem is, of course, equally (or more) important if you are both problem-stater and problem-solver. Difficulty in isolating the problem is often due to the tendency to spend a minimum of effort on problem-definition in order to get to the important matter of solving it. Inadequately defining the problem is a tendency that is downright foolish on an important and extensive problem-solving task. A relatively small time spent in carefully isolating and defining the problem can be extremely valuable both in illuminating possible simple solutions and in ensuring that a great deal of effort is not spent only to find that the difficulties still exist—perhaps in even greater magnitude.

> **Exercise:** Think of a problem that is bothering you. State your problem in writing as concisely as you can. Can you think of alternative problem statements that might be causing

the difficulties you are experiencing? If so, write them down and conjecture about the possible differences in solutions that occur to you.

Three: Tendency to Delimit the Problem Area Too Closely

Just as it is sometimes difficult to isolate the problem properly, it is also difficult to avoid delimiting the problem too closely. (In other words, one should not impose too many constraints upon it.) The following common puzzle is an example of this tendency to delimit too closely.

Puzzle: Draw no more than four straight lines (without lifting the pencil from the paper) which will cross through all nine dots.

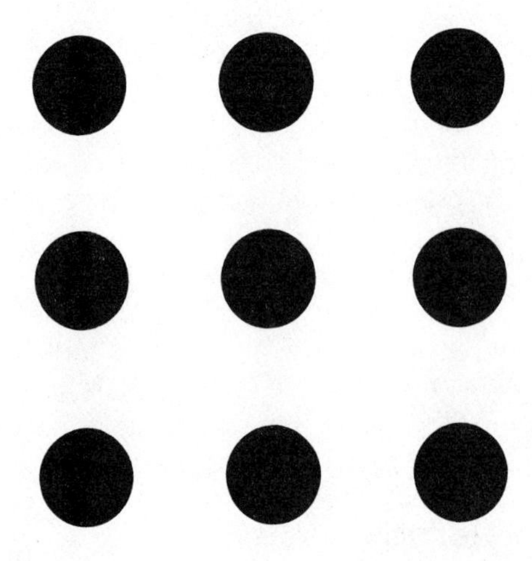

This puzzle is difficult to solve if the imaginary boundary (limit) enclosing the nine dots is not exceeded. One possible answer is shown below:

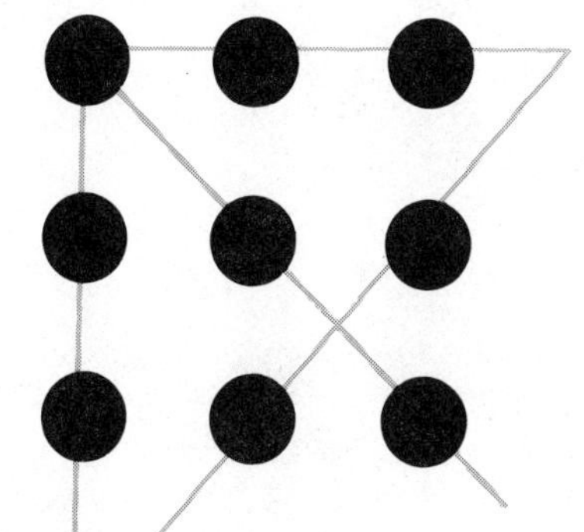

A surprising number of people will not exceed the imaginary boundary, for often this constraint is unconsciously in the mind of the problem-solver, even though it is not in the definition of the problem at all. The overly strict limits are a block in the mind of the solver. The widespread nature of this block is what makes this puzzle classic.

Such blocks are subtle and pervasive. For a talk which I once gave on the subject of problem-solving, an announcement was sent out with this puzzle on the cover. An anonymous party (confess) sent back the solution below:

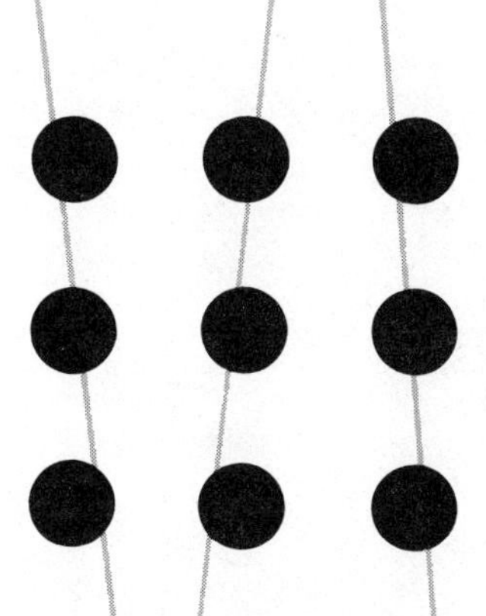

I officially designated him/her our official conceptual smart-ass and secretly admired that person because I was, of course, too blocked to realize that it wasn't necessary to draw the lines through the centers of the dots.

To add insult to this injury, one of my oldest friends later sent me the fiendish solution shown here, which allows all nine dots to be crossed off by one straight line—plus a little unblocked paper folding. Try this solution yourself—just cut out the adjoining page and start folding!

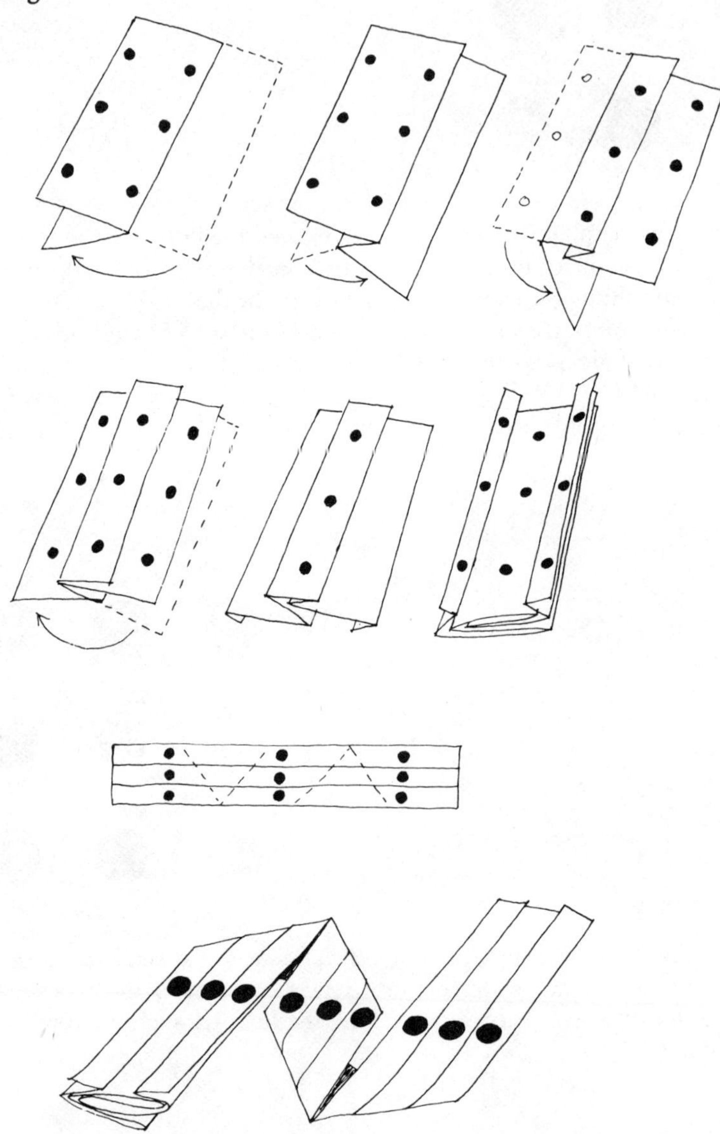

Fiendish Solution by Rodney W. Supple

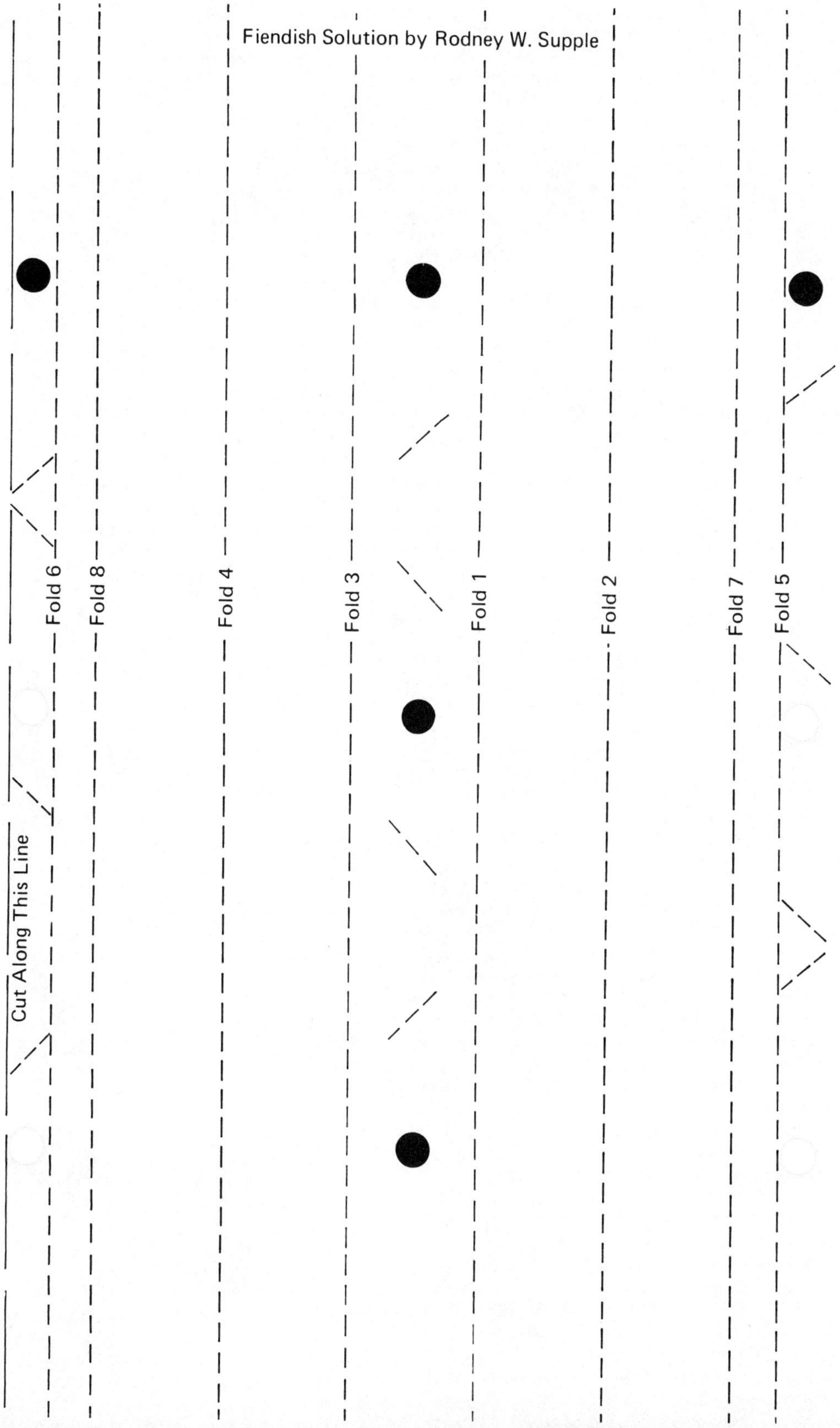

Cut Along This Line

Fold 6

Fold 8

Fold 4

Fold 3

Fold 1

Fold 2

Fold 7

Fold 5

I have received many such as the one below, which merely requires cutting the puzzle apart, taping it together in a different format, and again using one line.

It is also possible to roll up the puzzle and draw a spiral through the dots (below), and otherwise violate the two-dimensional format.

By now I have received dozens of answers to this puzzle, all exceedingly clever and all depressing in that I had thought of none of them.

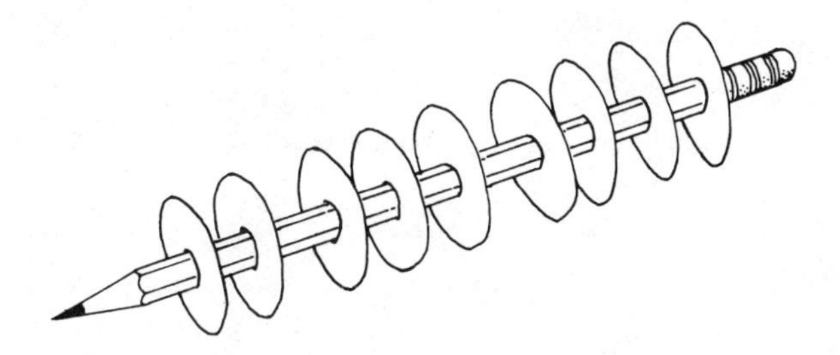

1 Line 0 Folds

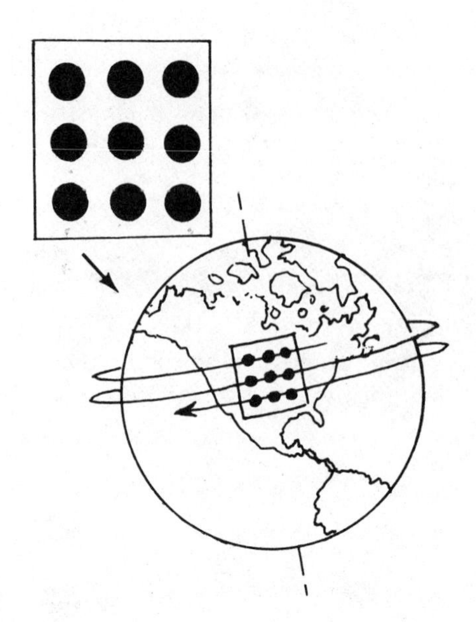

Lay the paper on the surface of the Earth. Circumnavigate the globe twice + a few inches, displacing a little each time so as to pass through the next row on each circuit as you "Go West, young man."

~ 2 Lines* 0 Folds

*Statistical

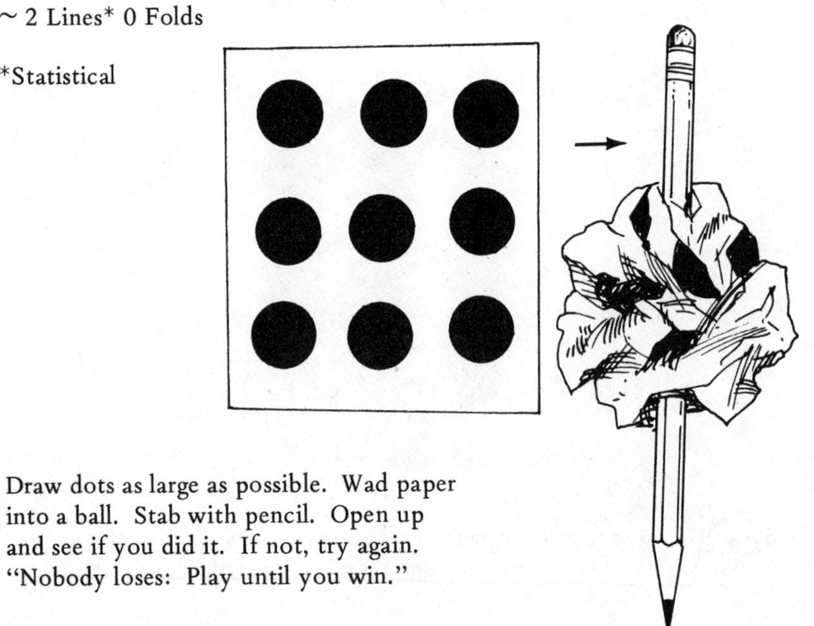

Draw dots as large as possible. Wad paper into a ball. Stab with pencil. Open up and see if you did it. If not, try again. "Nobody loses: Play until you win."

One of my favorites is reproduced on the next page.

May 30, 1974
5 FDR Navasa
Roosevelt Rds. ~~Na~~
Ceiba, P.R. 00635

Dear Prof. James L. Adams,
My dad and I were doing Puzzles from "Conceptual Blockbusting." We were mostly working on the dot ones, like
: : : My dad said a man found a way
. . .
to do it with one line. I tried and did it. Not with folding, but I used a fat line. I doesn't say you can't use a fat line.
Like this

Sincerely,
Becky Buechel
age: 10

P.S.
acctually you need a very fat writing apparatice

Outbursts of creativity such as these are exciting. One of the messages of this book is that we place limits upon our own functioning (the fence around the dots) and that once we realize the existence of these limits we will be eager to escape and will no longer be as hampered by them. The nine-dot puzzle is certainly evidence of this phenomenon. Limits are negotiable.

Just as a solution is sensitive to the proper isolation of the problem, it is also sensitive to proper delimitation (constraints). In general, the more broadly the problem can be stated, the more room is available for conceptualization. A request for the design of a better door will probably result in a rectangular slab with hinges and a handle. Is this what is wanted, or is the problem really a better way to get through a wall? A request for a better way to get through a wall releases one from the preconception of the rectangular slab that swings or slides. Students given this problem statement will come up with all manner of geometries for walls and openings, elastic diaphragms, mechanical shutters, curtains, and ingenious rotating and folding mechanisms. Is this what is wanted, or is the problem a better method of acoustical, visual, or environmental isolation? The solution of a laminar air curtain as is used to keep heat in stores or out of freezers while permitting free passage would not be likely to come from a harshly delivered "design a better door" problem statement.

If you hire an architect or a structural engineer or a lawyer, you are paying for his expertise. It is therefore foolish to constrain a problem so closely ("Here is a floor plan and an elevation—build it") that you are not taking advantage of the professional's abilities. This principle applies equally if a single person is both stating the problem and solving it. A problem statement which is too limited inhibits creative ability.

It is, of course, possible to err in the opposite direction and not delimit the problem sufficiently. The resulting solution may be so general or basic as not to be useful. An automobile company looking for a better way to keep windshields clean cannot do much with a solution that does away with automobiles. The proper statement of a problem therefore becomes a critical art, since it enables the extraction of the maximum of creative thought from the solver while still delivering a useful answer. I would hazard a guess, however, that more problems are overly limited in statement than inadequately limited. Because of that feeling and because this is a book on creativity, imposing too many constraints is expressly stated here to be *verboten*.

> **Exercise:** The next time you have a problem, solve it. Then, at your leisure, list at least three different possible delimitations of the problem and answers you might have come up with in each case. For instance, suppose that you are in your late thirties, your children are well into school, your husband is establishing a name for himself in his profession, and you are bored.

You might formulate your problem as "difficulty in establishing contact with the real world." You might contact several people you know and find a job working for a personable and rising young star in an interesting company. After a few exciting days in your new life, ask yourself how else you might have formulated your problem, and what might have happened. Perhaps:

1. You might have considered your problem to be "difficulty in dispelling boredom during the day" (a more delimited statement). You might have taken up several crafts, involved yourself in many lunches, committees, and volunteer activities, and attended many classes.

or:

2. You might have phrased your problem as "lack of a sufficiently challenging and productive career, now that the child-raising, home-establishing phase of life is under control" (a less delimited statement).

You might then have spent a large amount of effort outlining your goals and deciding how to best accomplish them and ended up as a law student.

or:

3. You might have decided that the problem is "role stereotyping that does not result in natural fulfillment for women" (even less delimited). You might have talked to people you know, professional people, people in educational institutions, and others and decided that a large-scale social problem exists. You might have then organized a city-wide, state-wide, or even national organization oriented toward helping women, such as you, to better enter the professional world.

As limits on problem-definition are relaxed, one usually becomes involved in interdisciplinary considerations: economic, political, and ethical. If you see your problem as simply "conforming with federal government smog regulations," your answer may be to put gadgets on existing engines. However, if you see your problem as "minimizing air pollution" you may consider entirely new concepts in transportation and will be involved in complex social as well as technical considerations.

Four: Inability to See the Problem from Various Viewpoints

It is often difficult to see a problem from the viewpoint of all of the interests and parties involved. However, consideration of such viewpoints not only leads to a "better" solution to the problem, in that it pleases more interests and individuals, but it is also extremely helpful in conceptualizing. Certainly in a problem between two people, the ability to see the problem from the other's point of view is extremely important in keeping the tone of the debate within reasonable bounds of refinement. In many cases, no solution is possible until each person can gain a feeling for the viewpoint of the other. Most problem solutions affect people other than the solver, and their interests must be considered. The architect must view the design of his buildings from the perspectives of his clients, his builders, suppliers, architectural critics, and others in his profession as well as from his own. The designer of an automobile should worry about those who must manufacture, operate, and maintain his output. The property owner building a fence must consider the viewpoints not only of his neighbors, the city council, his visitors, the garbageman, and passing motorists who can no longer see around the corner, but also of non-human participants such as his lawn, which may die in the shade of the fence, and the neighborhood cats, who may sit on his fence to better communicate their wails of war and love.

> **Exercise:** Think of an interpersonal problem you presently have. Write a concise statement of the problem as seen by each party involved. If possible show the statements to the corresponding parties and see if they agree with your interpretation of their perception of the problem.

In his book *New Think*, Edward de Bono talks about vertical and lateral thinking. Vertical thinking begins with a single concept and then proceeds with that concept until a solution is reached. Lateral thinking refers to thinking that generates alternative ways of seeing a problem before seeking a solution. At one point in his book, De Bono explains vertical and lateral thinking by referring to the digging of holes. He states:

> Logic is the tool that is used to dig holes deeper and bigger, to make them altogether better holes. But if the hole is in the wrong place, then no amount of improvement is going to put it in the right place. No matter how obvious this may seem to every digger, it is still easier to go on digging in the same place than to start all over again in a new place. Vertical thinking is digging the same hole deeper; lateral thinking is trying again elsewhere.

De Bono acknowledges advantages in digging in the same hole, admitting that "a half-dug hole offers a direction in which to expend effort." He elaborates, "No one is paid to sit around being capable of achievement. As there is no way of assessing such capability it is necessary to pay and promote according to visible achievements. Far better to dig the wrong hole (even one that is recognized as being wrong) to an impressive depth than to sit around wondering where to start digging." However, De Bono makes the point that many holes are being dug to an impractical depth, many in the wrong place, and that breakthroughs usually result from someone abandoning a partly-dug hole and beginning anew in a different place.

Five: Saturation

Saturation takes place with all sensory modes. If the mind recorded all inputs so that they were all consciously accessible, our conscious mind would be very full indeed. Many extremely familiar inputs are not recorded in a way that allows their simple recall.

Exercise: Without looking at one, draw an ordinary non-digital telephone dial, putting the letters and numbers in their proper locations.

Very few people can do this exercise successfully, even though they have used phone dials for a large proportion of their lives. However, the mind does not hold onto the locations of all the details of a phone dial, since it does not have to. If the letters and numbers were not marked on the phone, the mind would store the information for easy recall.

As other examples of saturation, you might try to draw (without looking at them) the grill on your car, your lawn mower handle, or any other object that you see repeatedly, but whose visual details are unimportant to you. Like a phone dial, even though you might think that you know the details, you cannot produce them when desired. The trickiest aspect of saturation is that you think you have the data, even though you are unable to produce it when needed.

Visual saturation is a problem in art schools, because it is necessary to teach students to see things they are used to ignoring. For this reason, beginning art students are sometimes told to do things like bending over and looking at the world upside down, since this upside-down orientation makes visible details that are usually not noticed (try it). Similarly if you look away from a nice sunset you notice all manner of usually unnoticed visual activity in the easterly direction, such as colors on clouds, muted tones on buildings, reflected lights on windows, etc.

Another situation that requires attention to saturation in problem-solving occurs when data arrives only occasionally or in the presence of large amounts of distracting data. Radar inputs in the military or in air traffic control are an example of this, as are irregularities in the operation of an airplane or ground vehicle that appear after a long period of normal behavior. The life of a professional pilot, for instance, has occasionally been described as years of tedium interspersed with seconds of terror. When the information resulting in this terror becomes available, it is obviously extremely important that the pilot notice it as soon as possible. Fortunately for us passenger-types, a great amount of effort on the part of human-factors engineers, psychologists, and equipment designers goes into ensuring that the tedium will be suitably interrupted.

Six: Failure to Utilize all Sensory Inputs

The senses are interconnected in a fairly direct manner. This will be discussed further in Chapter Six. Senses such as sight and hearing and taste and smell are commonly linked. Taste is severely inhibited if smell is suppressed. Similarly sight is augmented in a major way by sound (motion pictures).

Various sensory inputs (notably vision) are important to people who are extremely innovative. This is amply recorded in the literature. In a letter to Jacques Hadamard (taken from *The Creative Process*, edited by Brewster Ghiselin), Albert Einstein said: "The words or the language, as they are written or spoken, do not seem to play any role in my mechanism of thought. The psychical entities which seem to serve as elements in thought are certain signs and more or less clear images which can be 'voluntarily' reproduced and combined. . . . The above mentioned elements are, in my case, of visual and some of muscular type. Conventional words or other signs have to be sought for laboriously only in a secondary stage, when the mentioned associative play is sufficiently established and can be reproduced at will." Tesla, an extremely productive technological innovator (fluorescent lights, the A.C. generator, the "Tesla" coil), apparently had incredible visualization powers. As described by J.J. O'Neill in *Prodigal Genius: The Life of Nikola Tesla*, it was claimed that Tesla "could project before his eyes a picture complete in every detail, of every part of the machine. These pictures were more vivid than any blueprint." Further, Tesla claimed to be able to test his devices mentally, by having them run for weeks—after which time he would examine them thoroughly for signs of wear.

Problem-solvers need all the help they can get. They should therefore be careful not to neglect any sensory inputs. An engineer working on an acoustics problem for a concert hall, for instance, should not get so carried away with his theoretical analysis that he neglects to look at a wide variety of concert halls and listen to the quality of sound in each. He must also be aware that his acoustical treatment, although successful to the ear, may overly offend the eye and, if his material choice is extreme enough, perhaps also the nose.

It is for this reason that designers sometimes will consciously deprive themselves temporarily of certain sensory inputs to make sure they have adequately recorded others. The designer of a patio cover intended to take the place of shade trees until they grow high enough in a new yard would be well advised not only to look at trees, but also to listen to them, feel them, smell them, climb them, and generally saturate himself with them for a good bit of time before starting his design. In a well-working marriage, one is sensitive not only to the appearance of one's partner, but also to the sound, smell, taste, and feel of him or her. Problems between the two are best solved by utilizing inputs from all of these senses.

Convincing students that they should use all of the sensory inputs at their disposal is one of our most difficult tasks at Stanford. University students are highly verbal (they are admitted to school that way) and are relatively less competent visually. They are not used to relying on taste, smell, or feel for problem-solving. Generally speaking, they are familiar with problems that can be solved (they think) verbally or mathematically. They are not used to using sensory imagery in their thinking. This subject will be covered in more detail in Chapter Six, so we will dwell no more on it here. Suffice it to say that failure fully to utilize inputs from all the senses is a conceptual block that is quite common in problem-solving.

PARADE AMOUREUSE
Francis Picabia 1917

CHAPTER THREE

EMOTIONAL BLOCKS

THIS CHAPTER WILL BEGIN with a game—a game that requires a group of people, the larger the better, so try it at a party. It was, I think, invented by Bob McKim and is called "Barnyard."

> **Exercise:** Divide your group and assign them to be various animals as follows:
>
> | If their last names begin with: | they are: |
> |---|---|
> | A-E | sheep |
> | F-K | pigs |
> | L-R | cows |
> | S-Z | turkeys |
>
> Now tell each person to find a partner (preferably someone he does not know too well) and to look this partner in the eye. You will then count to three at which time everyone is to make the sound of his animal as loudly as he possibly can. See how loud a barnyard you can build.
>
> The participants in this game will be able to experience a common emotional block to conceptualization—namely, that of feeling like an ass. If you did not play the game and want to experience the feeling, merely stand alone on any busy corner (or wherever you are right now) and loudly make the sound of one of the animals.

As we will see in the next chapter, conceptualization is risky and new ideas are hard to evaluate. The expression of a new idea, and especially the process of trying to convince someone else it has value, sometimes makes you feel like an ass, since you are doing something that possibly exposes your imperfections. In order to avoid this feeling, people will often avoid conceptualization, or at least avoid publicizing the output.

Before we discuss specific emotional blocks, let me make a few comments about psychological theory. Although, as I stated earlier, psychological theory does not offer a complete model for explaining the conceptual process, many theories exist and have commonalities which are pertinent to understanding emotional blocks. Of particular importance are the theories of Freud and his followers and of the contemporary humanistic psychologists (Rogers, Maslow et al.).

Freud

Much of Freudian theory is based upon conflicts between the *id* (the instinctive animal part of ourselves) and the *ego* (the socially aware and conscious aspect) and *superego* (the moralistic portion of ourselves that forbids and prohibits). The motive force in the Freudian model is the id, which resides in the unconscious and is concerned with satisfying our needs. According to Freud, ideas originating in the unconscious must be subjected to the scrutiny of the ego (which may reject them be-

cause we cannot realistically carry them out) and the superego (which may reject them because we should not have let ourselves have such ideas in the first place). If these ideas are rejected, they will either be completely repressed or they will contribute to neurotic behavior because of unresolved conflict. If they are accepted, they will be admitted to the conscious mind. (This acceptance may be accompanied by anxiety, since once the ego and superego identify with an idea one can be hurt by its rejection.) If the ego and superego are overly selective, relatively few creative ideas will reach the conscious mind. If they are not selective enough, a torrent of highly innovative but extremely impractical ideas will emerge.

Since the time of Freud, his theory has been elaborated upon by his followers. A good example of this can be seen in Lawrence S. Kubie's book *Neurotic Distortion of the Creative Process.* Kubie utilizes the Freudian concept of *preconscious* in his model of creative thinking. He relegates the subconscious portions of creative thought and problem-solving to this preconscious, reserving the unconscious for unsettled conflicts and repressed impulses. In this model, the preconscious mental processes are hindered both by the conscious and the unconscious processes. As Kubie states in *Neurotic Distortion:*

> Preconscious processes are assailed from both sides. From one side they are nagged and prodded into rigid and distorted symbols by unconscious drives which are oriented away from reality and which consist of rigid compromise formations, lacking in fluid inventiveness. From the other side they are driven by literal conscious purpose, checked and corrected by conscious retrospective critique.

Like Freud, Kubie has a model of the mind in which creative thinking is inhibited by the conscious ego and superego and in which creativity occurs at least partly below the conscious level. However, neuroses play a much more villainous role in Kubie's model than in Freud's.

The Humanistic Psychologists

Although humanistic psychologists agree that creativity is a response to basic inner needs in people, they have a somewhat broader hierarchy of needs than the Freudians. They maintain that people create in order to grow and to fulfill themselves, as well as to solve conflicts and to answer the cravings of the id. They are more concerned with reaching upward and outward. Carl Rogers, in an article entitled "Toward a Theory of Creativity" in *Creativity and its Cultivation* (edited by Harold Anderson) explains:

> The mainspring of creativity appears to be the same tendency which we discover so deeply as the curative force in psychotherapy—man's tendency to actualize himself, to become his potentialities. By this I mean the directional trend which is evident in all organic and human life—the urge to expand, extend, develop, mature—the tendency to express and activate all the capacities of the organism, to the extent that such activation enhances the organism or the self. This tendency may become deeply buried under layer after layer of encrusted psychological defenses; it may be hidden behind elaborate facades which deny its existence; it is my belief, however, based on my experience, that it exists in every individual and awaits only the proper conditions to be released and expressed.

The humanistic psychologists feel that the creative person is emotionally healthy and sensitive both to the needs and the capabilities of his unconscious to produce creative ideas. Like Freud's creative person, he possesses a strong ego and a realistic superego which allow him to be a prolific conceptualizer and relatively free of distracting neuroses.

We can now come to several interesting and believable conclusions, based upon our brief discussion of psychology:

1. Man creates for reasons of inner drive, whether it be for purposes of conflict resolution, self-fulfillment, or both. He can, of course, also create for other reasons, such as money.
2. At least part of creativity occurs in a part of the mind which is below the conscious level.
3. Although creativity and neuroses may stem from the same source, creativity tends to flow best in the absence of neuroses.
4. The conscious mind, or ego, is a control valve on creativity.
5. Creativity can provoke anxieties.

Now I will continue with our discussion of emotional blocks.

Emotional blocks may interfere with the freedom with which we explore and manipulate ideas, with our ability to conceptualize fluently and flexibly—and prevent us from communicating ideas to others in a manner which will gain them acceptance. Let me list a few of them, which I will then discuss:

1. Fear to make a mistake, to fail, to risk
2. Inability to tolerate ambiguity; overriding desires for security, order; "no appetite for chaos"

3. Preference for judging ideas, rather than generating them
4. Inability to relax, incubate, and "sleep on it"
5. Lack of challenge (problem fails to engage interest) versus excessive zeal (overmotivation to succeed quickly)
6. Inability to distinguish reality from fantasy

Fear of Taking a Risk

Fear to make a mistake, to fail, or to take a risk is perhaps the most general and common emotional block. Most of us have grown up rewarded when we produce the "right" answer and punished if we make a mistake. When we fail we are made to realize that we have let others down (usually someone we love). Similarly we are taught to live safely (a bird in the hand is worth two in the bush, a penny saved is a penny earned) and avoid risk whenever possible. Obviously, when you produce and try to sell a creative idea you are taking a risk: of making a mistake, failing, making an ass of yourself, losing money, hurting yourself, or whatever.

This type of fear is to a certain extent realistic. Something new is usually a threat to the status quo, and is therefore resisted with appropriate pressure upon its creator. The risks involved with innovation often can result in real hardship. Far be it from me to suggest that people should not be realistic in assessing the costs of creativity. For instance, I spend a great amount of time attempting to explain to students that somehow the process of making money out of a commercially practical idea seems to require at least eight years, quite a bit of physical and emotional degradation, and often the sacrifice of such things as marriages and food. However, as I also try to explain to students, the fears that inhibit conceptualization are often *not* based upon a realistic assumption of the consequences. Certainly, a slightly "far-out" idea submitted as an answer to a class assignment is not going to cost the originator his life, his marriage, or even financial ruin. The only possible difficulty would arise if I, the teacher, were annoyed with his answer (and I happen to like such responses from students). The fear involved here is a more generalized fear of taking a chance.

One of the better ways of overcoming such a block is to realistically assess the possible negative consequences of an idea. As is sometimes asked, "What are your catastrophic expectations?" If you have an idea for a better bicycle lock and are considering quitting a job and founding a small business based upon the lock and a not-yet-conceived product

line to go with it, the risks are considerable (unless you happen to have large sums of money and important commercial contacts). If you invent a new method of flight (say, wings of feathers held together with wax) the risks may also be considerable in perfecting the product. However, if you think of a new way to schedule your day, paint your bathroom, or relate to others in your dormitory, the risks are considerably less.

In my experience, people do not often realistically assess the probable consequences of a creative act. Either they blithely ignore any consequences, or their general fear of failure causes them to attach excessive importance to any "mistake," no matter how minor it will appear in the eyes of future historians. Often the potential negative consequences of exposing a creative idea can be easily endured. If you have an idea that seems risky, it is well worth the time to do a brief study of the possible consequences. During the study, you should include "catastrophic expectations" (assume everything goes badly) and look at the result. By doing this, it will become apparent whether you want to take the risk or not.

> **Exercise:** Next time you are having difficulty deciding whether to push a "creative" idea, write a short (two-page) "catastrophic expectations" report. In it detail as well as you can precisely what would happen to you *if everything went wrong.* By making such information explicit and facing it, you swap your analytical capability for your fear of failure—a good trade.

No Appetite for Chaos

The fear of making a mistake is, of course, rooted in insecurity, which most people suffer from to some extent. Such insecurities are also responsible for the next emotional block, the "Inability to tolerate ambiguity; overriding desire for order; 'no appetite for chaos.' " Once again, some element of this block is rational. I am not suggesting that in order to be creative you should shun order and live in a totally chaotic situation. I am talking more of an excessive fondness for order in all things. The solution of a complex problem is a messy process. Rigorous and logical techniques are often necessary, but not sufficient. You must usually wallow in misleading and ill-fitting data, hazy and difficult-to-test concepts, opinions, values, and other such untidy quantities. In a sense, problem-solving is *bringing order to chaos.* A desire for order is therefore necessary. However, the ability to tolerate chaos is a must.

We all know compulsive people, those who must have everything always in its place and who become quite upset if the order of their physical lives is violated. If this trait carries over into a person's mental process, he is severely impaired in his ability to work with certain types of problems. One reason for extreme ordering of the physical environment is efficiency. Another may be the aesthetic satisfaction of precise physical relationships. However, another reason is insecurity. If your underwear is precisely folded and "dressed right," you have precise control over your underwear, and thus there is one less thing out of control to be threatening. I do not actually care how your underwear is stored. However, if your thoughts are precisely folded and dressed right you are probably a fairly limited problem-solver. The process of bringing widely disparate thoughts together cannot work too well because your mind is not going to allow widely disparate thoughts to coexist long enough to combine.

Judging Rather than Generating Ideas

The next emotional block, the "Preference for judging ideas, rather than generating them," is also the "safe" way to go. Judgment, criticism, tough-mindedness, and practicality are of course essential in problem-solving. However, if applied too early or too indiscriminately in the problem-solving process, they are extremely detrimental to conceptualization. In problem-solving, analysis, judgment, and synthesis are three distinct types of thinking. In *analysis,* there is usually a right answer. I am an engineer: if you pay me to tell you how large a beam is needed to hold up a patio roof, you rightly expect *the* answer. Fortunately, I know how to analyze such things mathematically and can give it to you. *Judgment* is generally used in a problem where there are several answers and one must be chosen. A court case is a good example. A situation such as Watergate is another. Judgments are made by sensible people as to guilt or innocence, and the situation is sufficiently complex that disagreements can occur. *Synthesis* is even more of a multianswer situation. A design problem (design a better way to serve ice cream) has an infinitude of answers, and there are few rigorous techniques to help in deciding between them.

If you analyze or judge too early in the problem-solving process, you will reject many ideas. This is detrimental for two reasons. First of all, newly formed ideas are fragile and imperfect—they need time to mature and acquire the detail needed to make them believable. Secondly, as we will discuss later, ideas often lead to other ideas. Many techniques of conceptualization, such as brainstorming, depend for their effectiveness on maintaining "way-out" ideas long enough to let them mature and spawn other more realistic ideas. It is sometimes difficult to hold onto such ideas because people generally do not want to be suspected of harboring impractical thoughts. However, in conceptualization one should not judge too quickly.

The judgment of ideas, unfortunately, is an extremely popular and rewarded pastime. One finds more newspaper space devoted to judgment (critic columns, political analyses, editorials, etc.) than to the *creation* of ideas. In the university, much scholarship is devoted to judgment, rather than creativity. One finds that people who heap negative criticism upon all ideas they encounter are often heralded for their practical sense and sophistication. Bad-mouthing everyone else's concepts is in fact a cheap way to attempt to demonstrate your own mental superiority.

If you are a professional idea-haver, your criticism tends to be somewhat more friendly. Professional designers are often much more receptive to the ideas of our students than non-design oriented faculty members. Professional problem-solvers have a working understanding of the difficulty in having ideas and a respect for ideas, even if they are flawed. If you are a compulsive idea-judger you should realize that this is a habit that may exclude ideas from your own mind before they have had time to bear fruit. You are taking little risk (unless you are excluding ideas that could benefit you) and are perhaps feeding your ego somewhat with the thrill of being able to judge the outputs of others, but you are sacrificing some of your own creative potential.

Inability to Incubate

The "inability to relax, incubate, and 'sleep on it' " is also a somewhat common emotional block. There is general agreement that the unconscious plays an extremely important role in problem-solving. Everyone has had the experience of having the answer to a problem suddenly occur in his mind. One maddeningly familiar phenomenon to many people is a late answer to an important problem. You may work for days or weeks on a problem, complete it, and go on to other activities. Then, at some seemingly random point in time, a better answer "appears." Since the original problem was probably completed in order to reach a deadline, this "better" answer often only serves to annoy you that you did not think of it sooner. This better answer came straight from the unconscious as a result of the "incubation" process it was going through. I have found in my own case that this "incubation" process works and is reliable. I have the confidence to think hard about a problem (charging up my unconscious) and then forget about it for a period of time. When I begin work on it again, new answers are usually present.

Many "symptoms" of incubation are common. There is a widespread belief among students that they do their best work just before deadlines. If, in fact, they work on the material when they receive it long enough to store the data in their unconscious, then incubation can occur, and a better solution may emerge at a later time. Incubation does often seem to produce the right answer at the appropriate time. Students often claim to have come up with a winning idea the morning that it is due, after struggling futilely with the problem for days.

You must allow the unconscious to struggle with problems. Incubation is important in problem-solving. It is poor planning not to allow adequate time for incubation in the solution of an important problem. It is also important to be able to relax in the midst of problem-solving. Your overall compulsiveness is less fanatical when you are relaxed, and the mind is more likely to deal with seemingly "silly" combinations of thoughts. If you are never relaxed, your mind is usually on guard against non-serious activities, with resulting difficulties in the type of thinking necessary for fluent and flexible conceptualization.

Lack of Challenge versus Excessive Zeal

"Lack of challenge" and "excessive zeal" are opposite villains. You cannot do your best on a problem unless you are motivated. Professional problem-solvers learn to be motivated somewhat by money and future work that may come their way if they succeed. However, challenge must be present for at least some of the time, or the process ceases to be rewarding. On the other hand, an excessive motivation to succeed, especially to succeed quickly, can inhibit the creative process. The tortoise-and-the-hare phenomenon is often apparent in problem-solving. The person who thinks up the simple elegant solution, although he may take longer in doing so, often wins. As in the race, the tortoise depends upon an inconsistent performance from the rabbit. And if the rabbit spends so little time on conceptualization that he merely chooses the first answers that occur, such inconsistency is almost guaranteed.

Reality and Fantasy

"Lack of access to areas of imagination," "Lack of imaginative control," and "Inability to distinguish reality from fantasy" will be discussed

The creative person needs to be able to control his imagination and needs complete access to it. If all senses are not represented (not only sight, but also sound, smell, taste, and touch) his imagination cannot serve him as well as it otherwise could. All senses need representation not only because problems involving all senses can be attacked, but also because imagery is more powerful if they are all called upon. If you think purely verbally, for instance, there will be little imagery available for the solving of problems concerning shapes and forms. If visual imagery is also present, the imagination will be much more useful, but still not as potent as if the other senses are also present. You can usually imagine a ball park much more vividly if you are able to recall the smell of the grass, the taste of the peanuts and beer, the feel of the seats and the sunshine, and the sounds of the crowd.

The creative person must be able not only to vividly form complete images, but also to manipulate them. Creativity requires the *manipulation* and *recombination* of experience. An imagination that cannot manipulate experience is limiting to the conceptualizer. You should be able to imagine a volcano being born in your ball park, or an airplane landing in it, or the ball park shrinking as the grass simultaneously turns purple, if you are to make maximum use of your imagination. Chapter Six will contain some exercises to allow you to gauge your ability to control your imagination as well as discussions on how to strengthen the "mental muscle" used in imagining.

The creative person needs the ability to fantasize freely and vividly, yet must be able to distinguish reality from fantasy. If his fantasies become too realistic, they may be less controllable. If you cannot go through the following exercise without a sense of acute physical discomfort, you may have difficulty distinguishing reality from fantasy. This exercise is taken from *Put Your Mother on the Ceiling* by Richard de Mille. Stay with each fantasy (marked off by slashes) until you have it fully formed in your imagination. This game is called *breathing*.

> Let us imagine that we have a goldfish in front of us. Have the fish swim around. / Have the fish swim into your mouth. / Take a deep breath and have the fish go down into your lungs, into your chest. / Have the fish swim around in there. / Let out your breath and have the fish swim out into the room again. /
>
> Now breathe in a lot of tiny goldfish. / Have them swim around in your chest. / Breathe them all out again. /
>
> Let's see what kind of things you can breathe in and out of

your chest. / Breathe in a lot of rose petals. / Breathe them out again. / Breathe in a lot of water. / Have it gurgling in your chest. / Breathe it out again. / Breathe in a lot of dry leaves. / Have them blowing around in your chest. / Breathe them out again. / Breathe in a lot of raindrops. / Have them pattering in your chest. / Breathe them out again. / Breathe in a lot of sand. / Have it blowing around in your chest. / Breathe it out again. / Breathe in a lot of little firecrackers. / Have them all popping in your chest. / Breathe out the smoke and bits of them that are left. / Breathe in a lot of little lions. / Have them all roaring in your chest. / Breathe them out again. /

Breathe in some fire. / Have it burning and crackling in your chest. / Breathe it out again. / Breathe in some logs of wood. / Set fire to them in your chest. / Have them roaring as they burn up. / Breathe out the smoke and ashes. /

Have a big tree in front of you. / Breathe fire on the tree and burn it all up. / Have an old castle in front of you. / Breathe fire on the castle and have it fall down. / Have an ocean in front of you. / Breathe fire on the ocean and dry it up. /

What would you like to breathe in now? / All right. / Now what? / All right. / What would you like to burn up by breathing fire on it? / All right. /

Be a fish. / Be in the ocean. / Breathe the water of the ocean, in and out. / How do you like that? / Be a bird. / Be high in the air. / Breathe the cold air, in and out. / How do you like that? / Be a camel. / Be on the desert. / Breathe the hot wind of the desert, in and out. / How does that feel? / Be an old-fashioned steam locomotive. / Breathe out steam and smoke all over everything. / How is that? / Be a stone. / Don't breathe. / How do you like that? / Be a boy (girl). / Breathe the air of this room, in and out. How do you like that?

It would certainly be uncomfortable to inhale sand. Whether you can imagine the feeling of inhaling sand depends somewhat upon your ability to fantasize. No danger exists from imagining such an act, and any pain felt is imagined, not real. However, if your fantasies are confused with reality, it can be very difficult to fantasize such things. The imagination is extremely powerful because it can go beyond reality. But in order to do this, the imagination must be set free of the constraints placed upon *real* acts and events.

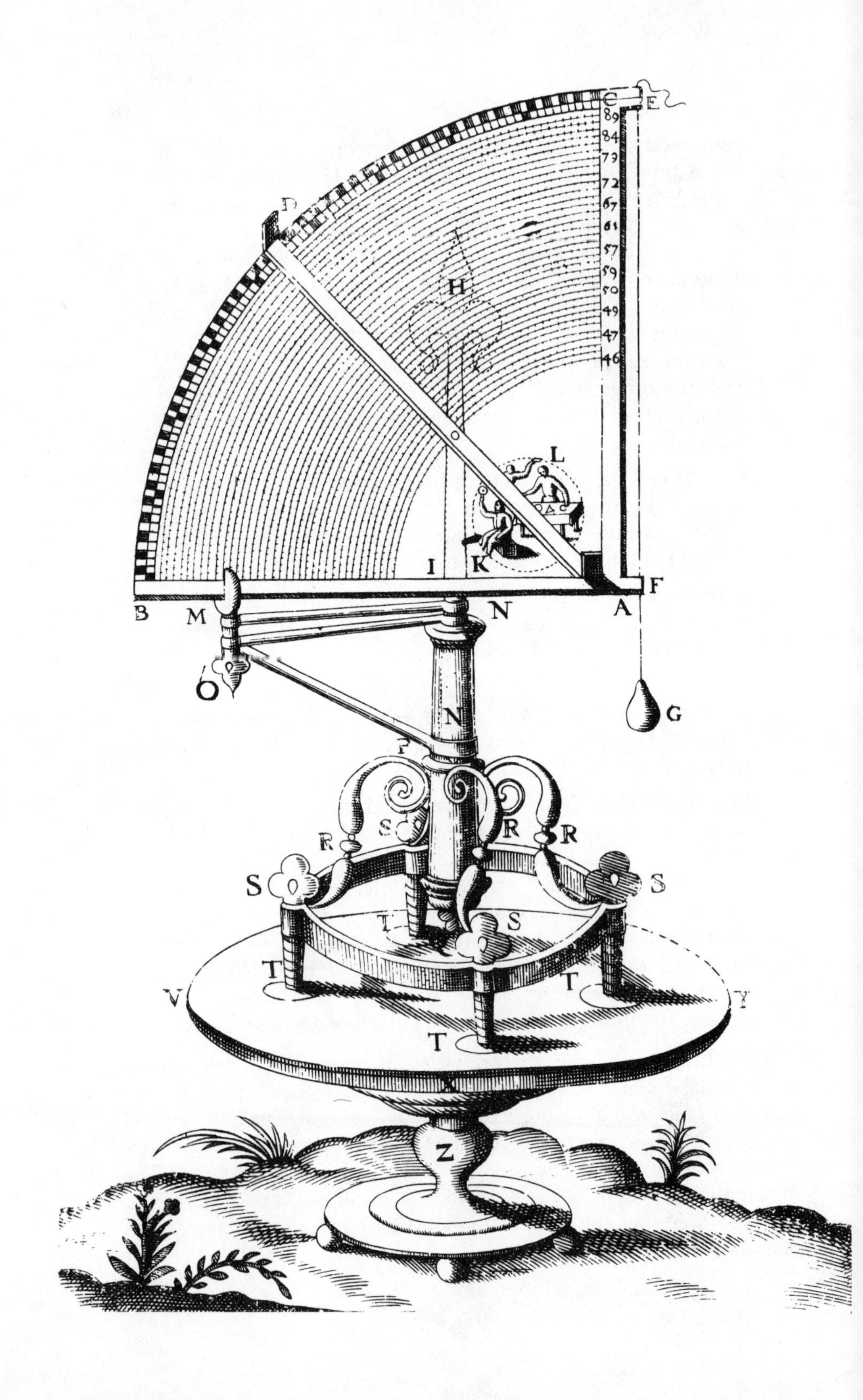
C
E
89
84
79
72
67
61
57
59
50
49
47
46
D
H
L
I
K
F
B
M
N
A
O
G
N
P
S
R
R
R
S
S
T
S
T
T
V
T
Y
Z

CHAPTER FOUR

CULTURAL AND ENVIRONMENTAL BLOCKS

CULTURAL BLOCKS ARE ACQUIRED by exposure to a given set of cultural patterns. Environmental blocks are imposed by our immediate social and physical environment. Since these two types of blocks are somewhat interrelated, we will discuss both of them in this chapter. Some examples of cultural blocks (for our culture) are:

1. Taboos
2. Fantasy and reflection are a waste of time, lazy, even crazy
3. Playfulness is for children only
4. Problem-solving is a serious business and humor is out of place
5. Reason, logic, numbers, utility, practicality are *good*; feeling, intuition, qualitative judgments, pleasure are *bad*
6. Tradition is preferable to change
7. Any problem can be solved by scientific thinking and lots of money

Some examples of environmental blocks are:

1. Lack of cooperation and trust among colleagues
2. Autocratic boss who values only his own ideas; does not reward others
3. Distractions—phone, easy intrusions
4. Lack of support to bring ideas into action

Let us discuss cultural blocks first. We will begin by working a problem that will make the message clearer.

> **Exercise:** Assume that a steel pipe is imbedded in the concrete floor of a bare room as shown below. The inside diameter is .06″ larger than the diameter of a ping-pong ball (1.50″) that is resting gently at the bottom of the pipe. You are one of a group of six people in the room, along with the following objects:
>
> 100′ of clothesline
> A carpenter's hammer
> A chisel
> A box of Wheaties
> A file
> A wire coat hanger
> A monkey wrench
> A light bulb
>
> List as many ways you can think of (in five minutes) to get the ball out of the pipe without damaging the ball, tube, or floor.

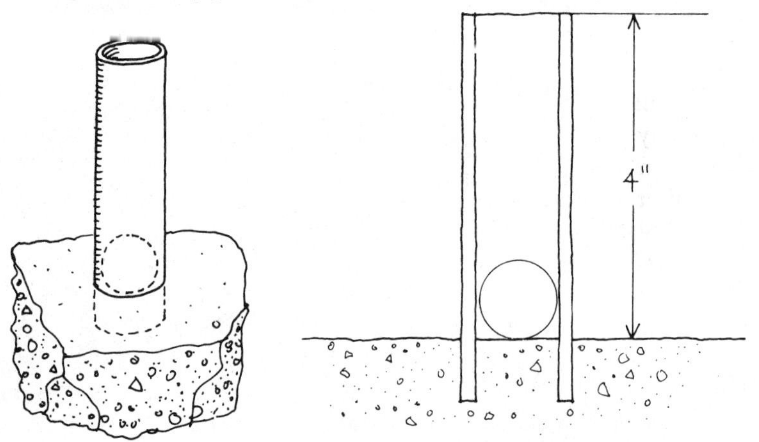

J.P. Guilford, one of the pioneers in the study of creativity, speaks a great deal about fluency and flexibility of thought. *Fluency* refers to the number of concepts one produces in a given length of time. If you are a fluent thinker, you have a long list of methods of retrieving the ball from the pipe. However, quantity is only part of the game. *Flexibility* refers to the diversity of the ideas generated. If you are a flexible thinker, you should have come up with a wide variety of methods. If you thought of filing the wire coat hanger in two, flattening the resulting ends, and making large tweezers to retrieve the ball, you came up with a solution to the problem, but a fairly common one. If you thought of smashing the handle of the hammer with the monkey wrench and using the resulting splinters to retrieve the ball, you were demonstrating a bit more flexibility of thought, since one does not usually think of using a tool as a source of splinters to do something with. If you managed to do something with the Wheaties you are an even more flexible thinker.

Did you think of having your group urinate in the pipe? If you did not think of this, why not? The answer is probably a cultural block, in this case a *taboo*, since urinating is somewhat of a closet activity in the U.S.

Taboos

I have used this ping-pong ball exercise with many groups and the response is not only a function of our culture, but also of the particular people in the group and the particular ambiance of the meeting. A mixed group newly convened in elegant surroundings will seldom think of urinating in the pipe. Even if members in the group do come up with this as a solution, they will keep very quiet about it. A group of people who work together, especially if all-male and if it's at the end of a working session, will instantly break into delighted chortles as they think of this and equally gross solutions. The importance of this answer is not that urinating in the pipe is necessarily the best of all solutions to the problem (although it is certainly a good one), but rather that cultural taboos can remove entire families of solutions from the ready grasp of the problem-solver. Taboos therefore are conceptual blocks. This is not a tirade against taboos. Taboos usually are directed against acts that would cause displeasure to certain members of a society. They therefore play a positive cultural role. However, it is the acts themselves which would offend. If imagined, rather than carried out, the acts are not harmful. Therefore, when working on problems within the privacy of your own mind, you do not have to be concerned with the violation of taboos.

Let us discuss a few more cultural blocks. The first two listed earlier, "Fantasy and reflection are a waste of time, lazy, even crazy" and "Playfulness is for children only," are challenged by quite a bit of evidence to indicate that fantasy, reflection, and mental playfulness are essential to good conceptualization. These are properties that seem to exist in children, and then unfortunately are to some extent socialized out of people in our culture. A four-year-old who amuses himself with an imaginary friend, with whom he shares his experiences and communicates, is cute. A 30-year-old with a similar imaginary friend is something else again.

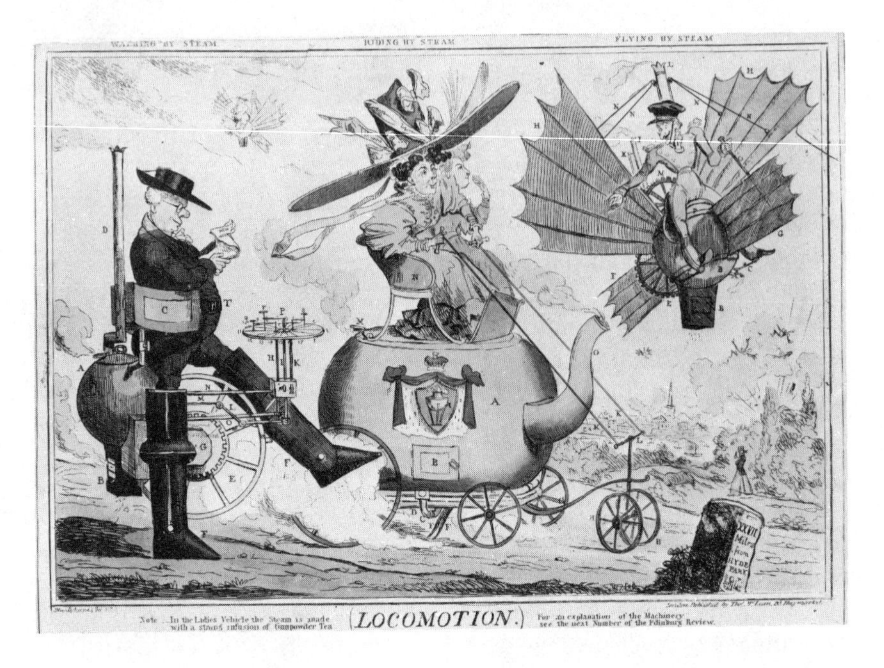

"Daydreaming" or "woolgathering" is considered to be a symptom of an unproductive person.

As mentioned previously, environmental and cultural blocks are somewhat interrelated. People can fantasize much more easily in a supportive environment. We quite frequently ask students to fantasize as part of a design task, and when assigned the task they do quite well. However, they tend to feel quite guilty if they spend their time in fantasy if it is not an assigned part of the problem, since it often seems to be a diversion. Nevertheless, if you are attempting to solve a problem having to do with bickering children, is it not worth the time and effort to fantasize a situation in which your children do not bicker and proceed to examine the situation closely to see how it works? If you are designing a new recreational vehicle, should you not fantasize what it would be like to use that vehicle?

Many psychologists have concluded that children are more creative than adults. One explanation for this is that the adult is so much more aware of practical constraints. Another explanation, which I believe, is that our culture trains mental playfulness, fantasy, and reflectiveness out of people by placing more stress on the value of channeled mental activities. We spend more time attempting to derive a better world directly from what we have than in imagining a better world and what it would be. Both are important.

Humor in Problem-Solving

Another cultural block mentioned was, "Problem-solving is a serious business and humor is out of place." In an essay, "The Three Domains of Creativity," Arthur Koestler, one of the more important writers who treat conceptualization, identifies these "domains" as *artistic originality* (which he calls the "ah!" reaction), *scientific discovery* (the "aha!" reaction), and *comic inspiration* (the "haha!" reaction). He defines creative acts as *the combination of previously unrelated structures in such a way that you get more out of the emergent whole than you have put in.* He explains comic inspiration, for example, as stemming from "the interaction of two mutually exclusive associative contexts." As in creative artistic and scientific acts, two ideas have to be brought together that are not ordinarily combined. This is one of the essentials of creative thinking. In the particular case of humor, according to Koestler, the interaction causes us "to perceive the situation in two self-consistent but habitually incompatible frames of reference." The joke-teller typically starts a logical chain of events. The punch line then sharply cuts across the chain with a totally unexpected line. The tension developed in the first line is therefore shown to be a put-on and with its release, the audience laughs. Let us look at a couple of jokes:

1. A man, on entering the waiting room of a veterinarian's office with his sick dog, sat next to a lady with a beautiful wolfhound. The wolfhound was extremely high-spirited and happily gamboled around the waiting room, as the man's own dog lay limply on the floor. Finally, curious as to why such an apparently healthy dog should be in a veterinarian's office, he turned to the lady and said:

   "You certainly have a beautiful dog."

   "Oh, thank you," she replied.

   "He looks so healthy," said he, "that I am surprised to see him in a veterinarian's office. What is wrong with him?"

   "Oh," she said with some embarrassment, "he has syphilis."

   "Syphilis!" he said. "How did he get syphilis?"

   "Well," she said, "he claims he got it from a tree."

   (attributed to Dorothy Parker)

2. A woman at a formal dinner was quite discomfited to observe that the man across from her was piling his sliced carrots carefully upon his head. She watched with horror as the pile grew higher and higher and the sauce began to drip from his

hair. She could finally stand it no longer, so she leaned toward him and said,

"Pardon me, sir, but why on earth are you piling your carrots on your head?"

"My God," said he, "are they carrots? I thought they were sweet potatoes."

(source unknown)

Is Koestler's explanation of comic inspiration correct? It would seem so in these two examples, since in each case a developing story (causing tension in the listener, who wants to know how it comes out) is smashed by another line of thinking which demonstrates to the listener that the whole thing is a farce. The listener then (we hope) laughs.

The critical point of interest here is that a similar reaction (laughter) may greet an original idea. A concept may be so contrary to the logical progress of the problem solution that it may be a tension release and cause laughter. Since an answer to a problem may release tension anyway, your unbelievably insightful solution to a problem may be greeted with giggles and hoots, not only from others but from yourself.

Creative groups with which I have been associated have been funny. So are creative people I have known. Humor is present in all manner of ways. I am not suggesting that creative activity is all fun, since it is fraught with frustration, detail work, and plain effort. However, humor is an essential ingredient of healthy conceptualization.

Reason and Intuition

The fifth cultural block on our list is "Reason, logic, numbers, utility, practicality are *good*; feeling, intuition, qualitative judgement, pleasure are *bad*." Reason, logic, numbers, utility, and practicality *are* good; but so, too, are feeling, intuition, qualitative judgment, and pleasure—especially if you are conceptualizing. This block against emotion, feeling, pleasure stems from our puritan heritage and our technology-based culture. It is extremely noticeable to me, since I work with large numbers of engineers and managers in situations where they must solve problems with a large amount of emotional content.

One cause for this block, which has complicated matters in the past but is hopefully dying out a little, has been the assigning of various mental activities and qualities to either the male or the female. In the past, it has been the female who was to be sensitive, emotional, appreciative of the fine arts, and intuitive. The male was to be tough, physical, pragmatic, logical, and professionally productive. Adhering to these constraints severely limits both sexes.

Abraham Maslow describes his findings about this block in his essay, "Emotional Blocks to Creativity" (found in *A Source Book for Creative Thinking*, edited by Parnes and Harding):

> One thing I haven't mentioned but have been interested in recently in my work with creative men (and uncreative men too) is the horrible fear of anything that the person himself would call "femininity," or "femaleness," which we immediately call "homosexual." If he's been brought up in a tough environment, "feminine" means practically everything that's creative. Imagination, fantasy, color, poetry, music, tenderness, languishing, and being romantic are walled off as dangerous to one's picture of one's own masculinity. Everything that's called "weak" tends to be repressed in the normal masculine adult adjustment. And many things are *called* weak which we are learning are not weak at all.

The opposite of this block also exists, of course. Many women are culturally conditioned to be as uncomfortable about many traits ascribed to the male (reason, logic, use of numbers, utility) as males are uncomfortable about "feminine" traits. Also, we find the current wave of anti-technology people who blame the technological emphasis in society for many of man's difficulties. These people believe that feeling, intuition, and qualitative judgment are good and that reason, logic, numbers, utility, and practicality are not all that exciting.

Effective conceptualization requires the problem-solver to be able to incorporate all of these characteristics—the use of reason and logic, as well as intuition and feeling. The designer of physical things must be aesthetically sensitive if the quality of our world is going to improve, whether the designer happens to be male or female. Similarly, the designer must be able to view technology honestly and without disciplinary bias whether from an art background or an engineering background. The businessman must use intuition and the social scientist must use mathematics. The man must be sensitive and the woman strong.

Left-Handed and Right-Handed Thinking

In reading the literature associated with conceptualization, one often encounters references to "left- and right-handed thinking." This is discussed particularly well by Jerome Bruner in his book, *On Knowing: Essays for the Left Hand*. The right hand has traditionally been linked with law, order, reason, logic, and mathematics—the left with beauty, sensitivity, playfulness, feeling, openness, subjectivity, and imagery. The right hand has been symbolic of tools, disciplines, and achievement—the left with imagination, intuition, and subconscious thinking. In Bruner's words:

> . . . the one the doer, the other the dreamer. The right is order and lawfulness, *le droit*. Its beauties are those of geometry and taut implication. Reaching for knowledge with the right hand is science. . . . Of the left hand we say that it is awkward. . . . The French speak of the illegitimate descendant as being *à main gauche*, and though the heart is virtually at the center of the thoracic cavity, we listen for it on the left. Sentiment, intuition, bastardy. And should we say that reaching for knowledge with the left hand is art?

Oddly enough, this historical symbolic alignment of the two hands with two distinct types of thinking is consistent with present understanding of brain function. The left hemisphere of the brain (which controls the right hand) contains the areas which are associated with control of speech and hearing and involved with analytical tasks such as solving an algebra problem. The right hemisphere (which controls the left hand) governs spatial perception, synthesis of ideas, and aesthetic appreciation of art or music. However, this coincidence is not the main message here, which is that the effective conceptualizer must be able to utilize both right-handed and left-handed thinking. C.P. Snow, in his famous book hypothesizing the existence of two cultures, *Two Cultures and the Scientific Revolution*, separates scientists from humanists. Yet, if one *can* separate people that clearly, then the people one has separated are not maximizing their creative potential. The scientists who are responsible for breakthroughs in knowledge cannot operate entirely by extrapolating past work, but must utilize intuition, too. Similarly, the humanists who disregard the logical are doomed to be ineffectual (even counterproductive) in influencing social actions.

An emphasis on either type of thinking—to the disregard of the other—is a cultural block. In the professional world in our culture, the emphasis is placed on right-handed thinking. It is easier to get money to support right-handed thinking than left-handed thinking. More fathers want their sons to be lawyers, doctors, or scientists than painters, poets, or musicians. Until the culture is willing to accept the equal importance of left- and right-handed thinking in both sexes, a large number of its members will continue to suffer from this conceptual block.

> **Exercise:** Put yourself into a left-handed thinking mode. Stay away from logic, order, mathematics, science. Think about your feelings, beauty, sadness, the inputs that are coming

to your senses. You can probably do this better by placing yourself in a conducive environment (under a tree in the springtime, alone in your most comfortable chair). Then switch yourself into a right-handed mode by thinking of a detailed plan to make money out of one of your left-handed thoughts. Are you ambidextrous? Are you able to shift from one type of thinking to the other, and ideally to do both at once? Or are you more comfortable with one type of thinking than the other?

The block entitled "Any problem can be solved by scientific thinking and lots of money" is of course a cultural one related to the emphasis on the importance of right-handed thinking. It is also interesting, because it exists partly as a result of popular misconception about the scientific process. Science depends both upon logical controlled progress (right-handed) and breakthroughs (often somewhat left-handed). Maslow, in his essay, "Emotional Blocks to Creativity," discusses primary creativity, which he describes as the "creativeness which comes out of the unconscious, and which is the source of new discovery (or real novelty) of ideas which depart from what exists at this point." This is the force behind the breakthroughs so necessary to science. He continues by speaking of what he calls secondary creativity, which he explains as follows:

> I am used now to thinking of two kinds of science, and two kinds of technology. Science can be defined, if you want to, as a technique whereby uncreative people can create and discover, by working along with a lot of other people, by standing upon the shoulders of people who have come before them, by being cautious and careful, and so on. That I'll call secondary creativeness and secondary science.

Primary and Secondary Creativity

The present awesome progress in genetics and biochemistry (through a large amount of secondary creativity) rests upon the discovery of RNA and DNA and their functions and structures (primary creativity). For a good treatment of this, read James P. Watson's *The Double Helix*, if you have not already. This is an intriguing book which talks about science in a way that is so contrary to many people's concept of the scientific method that it was very controversial when it first came out. It treats the discovery of the structure of DNA as a very human and very left-handed process. Watson and co-discoverer Francis Crick relied heavily on inspiration, iteration, and visualization. Even though they were superb biochemists, they had no precedent from which they could logically derive their structure and therefore relied heavily on left-handed thinking. The U.S. space effort during the 1960s was extremely impressive and exemplified the power of science—and technology based on science. However, a great deal of primary creativity and left-handed thinking was involved. Even such basic "scientific" decisions as whether to carry instruments to measure physical quantities or television cameras on the first lunar spacecraft were made in a left-handed way, since there was simply no way to make them with sheer logic. The design of the first spacecraft required a high degree of "art" (backed up, of course, by a great deal of analysis, detail design, and sophisticated fabrication and development) because there was no precedent that the designers could logically extend.

If "scientific thinking" is properly defined, it is extremely powerful for large-scale well-funded attacks on problems; however, right-handed science is only effective if based on established understanding. Right-handed science and lots of money can solve only problems that are solely in the domain of understood phenomena (a relatively small domain). Problems with social and emotional content and high complexity, such as crime in the cities, require a great deal more than right-handed science or secondary creativity.

Unfortunately, left-handed thinking and primary creativity are harder to explain, more difficult to predict, and less consistent than right-handed thinking and secondary creativity. It is therefore more difficult to write proposals that will bring support for such activities. It is easier for me to secure funding to work on the application of some newly discovered scientific phenomena (even though the potential good of the application may be small) than it is to find support for looking for a breakthrough. In the first case, the funding agency and I can be quite confident of the detailed nature of the work that needs to be done, the approximate amount of money needed, the schedule, and that I will in fact come up with something. In the second, there is no such security. The funding agency must judge me on the basis of intangibles such as my previous performance, my motivations, and my knowledge. The second is more of a gamble than the first. Support for science therefore also tends to be biased toward right-handed thinking, since most agencies handing out money must answer to someone and therefore tend to be somewhat conservative.

The "vagueness" of primary creativity and left-handed thinking, of course, also plagues those involved in the humanities and the soft sciences. Many of the soft sciences have sought to become more quantitative and rigorous in order to take better advantage of our cultural bias toward right-handed thinking. It is debatable whether this has been advantageous. Although a scientist, I am very sympathetic to the wails of those in the humanities and social sciences as to the lack of monetary support they receive from our society. At one point in my education (after I had become an engineer) I was enrolled in art school. A painting teacher I knew would from time to time tell me that I had an excellent background for painting. His reasoning was economic. He believed that many painters were hampered in the beginning of their careers by the necessity of holding down low-paying and long-houred jobs in order to support their families. He figured that I should be able to support myself by doing engineering work part-time and would therefore have time and energy available to paint. A strange observation, but perhaps a true one. It is much easier for me to find support for my life-style than it is for friends of mine who want to write or paint. The humanities and the social sciences are extremely vital in a mature society such as ours. Their importance is presently obscured by a massive cultural block.

Tradition and Change

As a final, subtle cultural block, I would like to discuss briefly the concept: "Tradition is preferable to change." In his book *Notes on the Synthesis of Form*, Christopher Alexander discusses two types of culture, one that he calls the unselfconscious culture, and one the self-conscious culture. The unselfconscious culture is tradition oriented. Traditional form and ceremonies are perpetuated, and often taboos and legends work against change. The architect in such a culture would probably serve a long apprenticeship and learn how to make the traditional buildings (the long house, the temple). When he reached a stage in which he was judged competent by his elders, he would presumably become a master and train other apprentices. The United States is hardly such a culture. Any young architect knows better than to study traditional building forms. Ours is a self-conscious culture. New religions, forms, social movements, and styles in dress, talk, entertainment, and living crop up continually. Age and experience are venerated only if "relevant," and long apprenticeships are rapidly becoming extinct. A very high value seems to be placed on innovation.

Yet, strangely enough, many individuals value tradition more than they do change. This is probably good, since in my opinion our culture has little enough tradition. However, as far as good conceptualization is concerned, such an attitude has negative effects. Motivation is essential to creativity. No matter how talented the problem-solver, frustration and detail work are inescapable in problem-solving. Unless you truly *want* to solve a problem (for pleasure, money, prestige, comfort, or whatever) you probably will not do a very good job. Unless you are convinced that change is needed in a particular area, you are not likely to hypothesize ways of accomplishing that change.

The problem arises when individuals become so universally in favor of tradition that they cannot see the need for and desirability of change in specific areas. The true conservative, I suppose, would fall in this category. Some environmentalists lose their credibility by being totally against change in an area. If a person is truly grounded in the "good old days," and feels strongly that changes in the past 20 or 30 years have diminished rather than enhanced the quality of life, he is unlikely to be *motivated* to be a very good conceptualizer. He is culturally blocked. The person who is in favor of change for change's sake may be a more dangerous animal to have around. Yet, as far as conceptualization (the subject of this book) is concerned, he is probably in fairly good shape.

Thinking Through Blocks

Projects requiring one to think through cultural blocks are among the most popular with our students, since the blocks are so difficult to overcome and yet so obvious once they have been overcome. We often ask our students to design puzzles, games, or situations for each other that require breaking through a cultural block in order to reach a solution. One project that sticks in my mind required that a dollar bill be removed from beneath a precariously balanced object without tipping over the object. This was extremely easy to do if the bill were torn in half. However, for various cultural reasons (it's illegal to deface money, one doesn't usually tear up things of value), no one thought of this particular solution, with the result that no one could remove the dollar. Another project required that one playing card out of a deck of 52 be destroyed. Once again no one thought of perpetrating such a crime (we are a society of card players and most of us do not approve of incomplete decks of cards). Still a third I can remember was perhaps the most basic I have seen. The solution of the problem required that a number of objects be moved around a board in a prearranged sequence in order to reach the desired final configuration. It turned out to be impossible to follow the rules and solve the problem. The cultural block? Following rules! It was simple to attain the desired configuration if the rules were violated.

A less flippant situation occurs when students from more rigid and theory-oriented disciplines take courses in design. Expertise in design is somewhat different from expertise in, say, fluid mechanics. Design is a multi-answer situation and analysis is used to gain an end, not for its own sake. The teacher, although hopefully experienced in the design process and in command of the necessary techniques, is not the usual type of academic expert, in that he or she does not have a monopoly on the "right" answers at the beginning of the course, and in fact may not even always come up with the "best" answer. Grading becomes much more subjective and the student must take more academic risk, since the evaluation standards are less orthodox. Students from a school system in which grading is extremely important, and in which the professor or teacher is an extreme authority figure, sometimes have difficulty in adapting to design courses. They are often preoccupied with "What is the answer?" and "How do I ensure that I will get an A?"—as well they should be, since their background has been exclusively oriented in such directions. The tragedy is that many foreign students from countries that need capable designers and problem-solvers suffer from such blocks. Academic risk-taking is somewhat of a taboo. Another culturally-induced difference between students from the U.S. and those from less industrially developed countries is the difference in their knowledge of, and attitude toward, machines. Students from the U.S., Western Europe, etc. have grown up with cars, motorcycles, and other such devices and are quite at home with them. Students from less industrially developed countries often have had less opportunity in their cultures to be exposed to machinery and are therefore somewhat less experienced and more inhibited in working with it.

Environmental Blocks

Let us now move on to environmental blocks. These are blocks that are imposed by our immediate social and physical environment. The most obvious blocks are the physical. Plainly the physical surroundings of the problem-solver influence his productivity. I am sure that all of you are familiar with the effect of distractions. It is very difficult to work on complicated problems with continual phone interruptions. At times even potential distractions are a problem since when you are in a frustrating phase of problem-solving, you are quite tempted to take advantage of such opportunities. Personally speaking, when involved in problem-solving I will go to heroic efforts to be distracted. Often I have to force myself out of bed at an inhuman hour in the morning to work on a problem when I am sure I can find no alternative activities available and no one to talk to. Even then, I often just sit hoping that someone will wake up and distract me.

The physical environment affects everyone. Yet, because of the individual habit patterns we all acquire, different individuals are affected differently. With regard to mental activity, some people work better in cold rooms, some in warm rooms, some in cold rooms with their feet wrapped in something warm. Some people work better to music and some in silence; some around others and some in isolation; some in windowless rooms and some in rooms with windows. Some are impervious to their visual surroundings and others are very sensitive to them.

Supportive Environments

In his book, *The Art and Science of Creativity*, George Kneller discusses some of the sometimes bizarre devices many writers have adopted with respect to their working environment: "Schiller, for example, filled his desk with rotten apples; Proust worked in a cork-lined room; Mozart took exercise; Dr. Johnson surrounded himself with a purring cat, orange peel, and tea; Hart Crane played jazz loud on a Victrola. All these are aids to the intense concentration required in creative thinking. An extreme case is Kant, who would work in bed at certain times of the day with the blankets arranged round him in a way he had invented himself. While writing *The Critique of Pure Reason* he would concentrate on a tower visible from his window. When some trees grew up to hide the tower, he became frustrated, and the authorities of Königsberg cut down the trees so that he could continue his work."

Some people may have a particular environment in which they are most effective at conceptual work of any kind. Therefore we sometimes find the all-purpose studio, in which a person may paint, write, sculpt, invent, and whatever. Another person may have one environment in which he can best write, another in which he can best throw pots, and still a third in which he does woodworking. Even though such individual differences exist, we can still say that most individuals do conceptual work best in a particular type of environment.

> **Exercise:** Take a piece of paper and list the characteristics of the most supportive possible environment you can think of for your own conceptual work (or different types of environment for different types of work). Do the environments in which you work resemble this? If not, why not? Assuming

your hypothesized environment is practical for you (not the beaches of an as yet undiscovered South Sea island), change your working environment to more closely resemble your hypothetical one. Does this make an appreciable difference on your conceptual productivity?

Although environment usually has physical connotations, the most important environmental blocks are often not physical. In fact, if anything they verge on the cultural and on the emotional. As discussed in the last chapter, conceptualization involves a certain amount of emotional risk. Change is often threatening; therefore, so are new ideas. They can be quickly squelched, especially when newly born, imperfect, and not reduced to practice. The usual response of society, in fact, *is* to squelch such ideas. There are many ways to do this. One is to over-analyze them. Another is to laugh at them. Still another is to ignore them.

Exercise: Think up a new idea, maybe an invention, that sounds reasonably plausible. Maybe an electric toilet brush, or a mail campaign to convince the post office to improve its service, or anything else. Then seriously propose this idea to friends and (if you are brave) others you meet from time to time. Note their reactions. Are any, other than your friends, enthusiastic? (Are even your friends really receptive, or are they merely being polite?) This is a poor experiment since some of your ideas may be brilliant and some terrible, and this conceivably could influence the response. However, I do not think that the difference in response will be that large. If you want to improve the experiment, try both a brilliant and a poor idea on the same people.

Accepting and Incorporating Criticism

Non-supportive responses are especially harmful when they come from bosses, colleagues, or friends. In Chapter Eight, we will discuss conceptualization in groups and in organizations. However, a few comments are in order here. An atmosphere of honesty, trust, and support is absolutely necessary if most people are to make the best of their conceptual abilities. There are exceptions, it is true. Many of the outstanding inventors I have known have been quite confident of their abilities and less dependent on support from others. One of the best of these idea-havers worked with me at one time. Given a problem he would instantly throw together a solution. These solutions were often so poorly thought out that I would almost break out in a rash. He would then happily go to the next office and receive enough criticism on the idea to send me into a depression for several days. He would then incorporate the criticism into his idea and proceed to the next office. In this way, he would literally construct a solution and usually an outstanding one. He was successful because of his ability to accept and incorporate criticism. However, people like this are rare.

Most people are not happy with criticism and, to make matters worse, are somewhat unsure of the quality of their own ideas. They therefore require a supportive environment in which to work. One of our most serious problems with students in design classes is that they hesitate to expose ideas about which they are unsure, not only to the faculty, but also to each other. Since many of their creative ideas fall into this not-sure category (naturally, since they have little else to judge these ideas by) they hesitate to reveal them. We have to convert the class (usually a listening, competitive, no-risk situation) into a friendly, noncompetitive, interactive situation in which people will take the risk of exposing their most impractical ideas to each other. Competition and lack of trust destroy such a supportive environment. No one likes to expose his magnificent concept if someone is going to steal it or be jealous.

Autocratic Bosses

Bosses with answers are a particular problem in the engineering profession. Many productive problem-solvers are strong-headed. They can carry a concept through to completion in spite of apathy or hostility from others and the difficulty of finding support for a new idea. If they happen to have good judgment, they are able to accomplish noticeable achievements in a company environment and are often promoted in management. One therefore often finds that many managers are successful idea-havers who are stubborn enough to push their ideas through to completion. They tend to continue in this mode when managing others. Although a manager such as this can be an effective problem-solver, he is essentially operating with his own conceptual ability and an in-house service organization—he is probably not going to make much use of the conceptual ability of his subordinates. In order to maximize the creative output of a group, a manager must be willing and able to encourage his subordinates to think conceptually and to reward them when they succeed. He should, of course, conceptualize on his own. But he should do it somewhat in tandem with the other members of his group, if he is attempting to use them to their fullest. This is an obvious piece of advice that is surprisingly often ignored. Time and time again I have seen design groups operating mainly on the concepts of the group leader. Such a group admittedly can be successful if the leader is an outstanding conceptualizer and the members of the group are content to develop his ideas. However, our concern is with environmental blocks, and such a working situation is hardly an environment conducive to conceptualization on the part of the group members.

Non-Support

Lack of physical, economic, or organizational support to bring ideas into action is also another common problem. New ideas are typically hard to bring into action. A great amount of effort is involved in perfecting an idea and then selling it. Many conceptual breakthroughs in science, for instance, have taken years of work to validate to the point where they would elicit interest from others in the scientific community. A novel itself is far removed from the original thought that inspired it. Even after the idea is fleshed out into a believable and complete form, it must be sold to an often skeptical world. This may require money and time. Again, using the inventor as an example: the small inventor is at a distinct disadvantage compared to the corporate inventor because of the fabrication support he may need, the test equipment he may desire, the legal and promotional expertise he may require, and the food and rent his family will consume while he is doing his inventing. Even the best of ideas is doomed if time and money are not available to push it to fruition.

Granted, the inventor is perhaps an extreme example. Nonetheless, even a concept for a new recipe is useless without the money to buy the ingredients and the time to cook it. A concept for a painting or a drawing is similarly useless without the supplies and the time. A concept for improving a marriage (take a vacation) requires economic and temporal support. All ideas require an environment that will produce the support necessary to bring them to fruition. This support may come from your friendly venture-capital firm, your bank, your spouse, your income surplus, or any other form of patronage. Lack of such patronage is a very effective environmental block.

VUE FROM SPOHRPLATZ - FRIEDRICHSPLATZ - "Documenta"
Christo
PROJET POUR MONUMENTAL EMPAQUETAGE POUR DOCUMENTA 1968

CHAPTER FIVE

INTELLECTUAL AND EXPRESSIVE BLOCKS

INTELLECTUAL BLOCKS RESULT in an inefficient choice of mental tactics or a shortage of intellectual ammunition. Expressive blocks inhibit your vital ability to communicate ideas—not only to others, but to yourself as well. Let us look at the following blocks:

1. Solving the problem using an incorrect language (verbal, mathematical, visual) — as in trying to solve a problem mathematically when it can more easily be accomplished visually
2. Inflexible or inadequate use of intellectual problem-solving strategies
3. Lack of, or incorrect, information
4. Inadequate language skill to express and record ideas (verbally, musically, visually, etc.)

A few examples should help us understand these blocks better. The monk puzzle described in the first chapter of this book is one in which choosing the correct language (visual) leads you rapidly toward a solution. Here is another "language" problem:

Exercise: Picture a large piece of paper, the thickness of this page. In your imagination, fold it once (now having two layers), fold it once more (now having four layers), and continue folding it over upon itself 50 times. How thick is the 50-times-folded paper?

It is true that it is impossible to fold any piece of paper, no matter how big or how thick, 50 times. But for the sake of the problem, imagine that you can. When you either have the answer or have given up, continue.

Your *first fold* would result in a stack 2 times the original thickness. Your *second* would give you a stack 2×2 times the original thickness. Your *third:* $2 \times 2 \times 2$ times the original thickness. Extending this, if you are somewhat of a mathematician, you should recognize that the answer to the problem is 2^{50} times the original thickness (2^{50} happens to be about 1,100,000,000,000,000). If the paper is originally the thickness of typing paper, the answer is some 50,000,000 miles or over half the distance from the earth to the sun.

If you tried to attack this particular problem with visual imagery (the clever way to handle the monk puzzle) you probably could not get an answer, since it is next to impossible to accurately visualize 50 folds. If you attacked it verbally, you probably also had trouble. If you are familiar with doubling problems, you knew that the answer was a surprisingly big number, but still could not place a value on it. The correct language in this problem was clearly mathematics.

Choosing Your Problem-Solving Language

Once again, how did you select the mental strategy you used to work on this problem? How did you decide to use visualization, mathematics, or whatever? If you were faked into visualization by our mention of the monk problem, you chose it consciously. If you are really getting the message of this book, you consciously thought about various ways of working the problem and then picked one. However, many of you probably once again *unconsciously* selected a strategy and then unconsciously switched from one strategy to the other. As we said before, most people follow this *habit* pattern in problem-solving. Without conscious thought, a direction will occur in the mind. This direction may or may not be the right one. If it is a wrong one, another may or may not appear.

It is possible to aid this strategy selection by consciously considering the various languages of thought you might use. For instance, you could have read the paper folding problem and then said to yourself, "Let's see, this guy has been trying to sell me visual thinking. Can I solve it visually? I'll try a few folds. [Task becomes difficult.] What else could I try? Verbalization? Probably not—since it is a physical problem asking for quantitative data. Hey, quantitative—how about mathematics?" At this point, you either solve it by inspection (you're a pro), write out equations and solve them (semi-pro) or ask someone you know who knows math (amateur).

Here is another puzzle. Before you try thinking of the answer, examine the problem and see what mental languages seem appropriate. Then attack the problem in the most appropriate language:

> **Exercise:** A man and a woman standing side by side begin walking so that their right feet hit the ground at the same time. The woman takes three steps for each two steps of the man. How many steps does the man take before their left feet simultaneously reach the ground?

This is a good problem to solve with visual thinking. A live experiment with another person, a drawing, or a musical rhythm analogy will all work well. A mathematical approach will work, although it is somewhat circuitous. Verbalization, once again, will not get you very far. What language did you pick? Did it work? Did you try alternate approaches? How did you know it was time to give up on one and try another? The answer is that their left feet never hit the ground simultaneously.

Choice of the proper problem-solving language is difficult not only because the choice is usually made unconsciously, but also because of the heavy emphasis on verbal thinking (with mathematical thinking a poor second) in our culture. The two problems you just worked were difficult because neither can be easily solved by the application of verbal thinking. Visualization, as expressed through the use of drawings, is almost essential in designing physical things well. One reason for this is that verbal thinking, when applied to the design of physical things, has the strange attribute of allowing you to think that you have an answer when, in fact, you do not. Verbal thinking among articulate persons is fraught with glib generalities. And in design it is not until one backs it up with the visual mode that he can see whether he is fooling himself or not.

One can also talk about the inflexible or inadequate use of problem-solving strategies as a type of awareness block. Interaction Associates, founded by David Straus and Michael Doyle, has for some time been concerned with the effective use of thinking strategies. Interaction Associates trains facilitators for problem-solving groups, offers educational programs, and conducts research in problem-solving. In one of their publications, *Summary of Basic Concepts,* they explain:

> Each physical action or operation that we make to solve a problem can be seen in terms of a more general conceptual approach, useful in solving any problem. It is the rationale or purpose behind your actions: the "why" as opposed to the "what." This general, conceptual approach we call a "strategy." In our terms, the concept inherent in a strategy is independent of context. In other words, a strategy should be able to be used in almost all kinds of problems. We find that the strategic level is one of the most useful ways of talking about problem-solving.

Flexibility in Your Use of Strategies

Interaction Associates believes strongly in the effectiveness of becoming aware of specific thinking strategies. They have worked with problem-solving groups in educational, business, and political settings. One of its major techniques with all groups is to keep track of the strategy or strategies being used at any time during a problem-solving session and to suggest changes or additions if the problem-solving process appears to be bogging down or overlooking possible approaches to solutions. In their *Strategy Notebook* they list some 66 strategies, accompanying each with a description of the strategy, a list of its advantages and disadvantages, and a sample exercise. A list of the strategies contained in *Strategy Notebook* is shown below, along with a sample page.

Build up
Eliminate
Work Forward
Work Backward
Associate
Classify
Generalize
Exemplify
Compare
Relate
Commit
Defer
Leap In
Hold Back
Focus
Release
Force
Relax
Dream
Imagine
Purge
Incubate

Display
Organize
List
Check
Diagram
Chart
Verbalize
Visualize
Memorize
Recall
Record
Retrieve
Search
Select
Plan
Predict
Assume
Question
Hypothesize
Guess
Define
Symbolize

Simulate
Test
Play
Manipulate
Copy
Interpret
Transform
Translate
Expand
Reduce
Exaggerate
Understate
Adapt
Substitute
Combine
Separate
Change
Vary
Cycle
Repeat
Systemize
Randomize

Eliminate

POWERS The power of elimination lies in the possibility that you may be more sure of what you don't want than what you do want. This strategy requires beginning with more than you need or want in the solution and eliminating elements according to some determined criteria. There is an element of safety in this strategy because you have not overly extended yourself by deciding what you don't want in the solution.

LIMITATIONS This strategy assumes that within the realm of possibilities you are considering, there is a good solution. However, after you've finished eliminating, it's possible to end up with nothing. Another difficulty is that it is easy to infer that you want the opposite of what you have eliminated (i.e., you don't want rain, therefore you must want sunshine, leaving out the possibilities of snow, fog, hail, etc.). Thus elimination must be tempered by caution and good judgment.

EXERCISE — I GOT RELIGION Have each member of your group build upon the subject of religion. Each member should offer any ideas or associations he has with the subject, and the ideas should be recorded. Once the group is satisfied that they have exhausted their resources, each member of the group should take a piece of paper and a pencil and review the recorded list, eliminating whatever they don't want included in their personal religion or philosophy, and writing down on their lists anything that is left. Once everyone has finished, pin the sheets of paper to a display board so that the members of the group can share each other's ideas. This exercise has the advantage of allowing the participants to get personally involved in the subject matter through use of the strategy of elimination. The exercise can also be modified to encompass a variety of subjects. This may prove to be an effective introductory experience for a humanities or comparative religion class.

Most people have no trouble in understanding such problem-solving strategies, once definitions and examples are made available. In fact, most people have *unconsciously* used all of them at one time or another. However, since the mind is used to selecting strategies subconsciously, it takes awareness of these strategies and *conscious* choice or an outside facilitator to make the best use of them on a specific problem. The Introduction to *Process Notebook*, also by Interaction Associates, summarizes the situation as follows:

> Just as we use physical tools for physical tasks, we employ conceptual tools for conceptual tasks. To familiarize yourself with a tool, you may experiment with it, test it in different situations, and evaluate its usefulness. The same method can be applied to conceptual tools. Our ability as thinkers is dependent on our range and skill with our own tools.

It is obvious that a compromise has to be reached in the conscious selection of thinking modes and problem-solving strategies. You should not devote 95 percent of your mental energy to the selection of strategies and thinking modes and reserve only 5 percent for the solving of the problem. Yet you should certainly spend some conscious effort thinking about strategies. First, by selecting strategies consciously you can often find approaches you would never have known about had you left the selection to your subconscious. Second, by becoming aware of various thinking strategies, what they can do, and how to use them, you can ensure that the mind has a larger selection when it utilizes its subconscious selection method. You can essentially become your own "facilitator."

Importance of Correct Information

Lack of, or incorrect, information is a third intellectual block. As we discussed earlier, Arthur Koestler in "The Three Domains of Creativity" states: "The creative act consists in combining previously unrelated structures in such a way that you get more out of the emergent whole than you have put in." Other definitions of creativity also emphasize this "combining" aspect. Plainly we must have the components to combine (information). But let us look at what happens if some of our components are incorrect. We will consider a situation in which each component appears no more than once and in which the order of combination is important.

If we combine two quantities, *a* and *b*, we have four possible results (*a*, *b*, *ab*, and *ba*). If *a* is incorrect, three of these results contain erroneous information. If both *a* and *b* are incorrect, all of them are contaminated. If we combine three quantities, *a*, *b*, and *c*, we have 15 possible results (*abc*, *acb*, *bac*, *bca*, *cab*, *cba*, *ab*, *ba*, *ac*, *ca*, *bc*, *cb*, *a*, *b*, *c*). If *a* is incorrect, 11 of these results contain erroneous information. If both *a* and *b* are incorrect, 14 of them are wrong. By playing with a little mathematics, we can come up with a general expression for this contamination tendency.

Let me first briefly refresh the memories of those who have studied math and perhaps enlighten those who have not. Let us assume that we have n elements. Then we see that there are a total of n possible arrangements (permutations) which contain only one element. There are $n(n-1)$ arrangements containing two elements, $n(n-1)(n-2)$ arrangements containing three elements, and so on, until we reach the number of possible arrangements containing *all* n elements. (There are $n!$, read n factorial, which is equal to $n(n-1)(n-2)\ldots(1)$, arrangements containing n elements.) Hence, the total number of arrangements (N) possible for n elements is the *sum* of the above terms. Or mathematically:

$$N = n + n(n-1) + n(n-1)(n-2) + \ldots + n!$$

For example, if we want to know how many arrangements are possible when we have 4 elements (a, b, c, d), the solution is:

$$N_4 = 4 + 4(3) + 4(3)(2) + 4(3)(2)(1) = 64$$

Now to continue. We can use this expression not only to calculate the number of possible arrangements of n elements (N), but also to find the number of them that are affected by erroneous quantities. If *one* quantity of the n is wrong, the number of arrangements which do *not* contain the erroneous element is simply the number of arrangements which can be formed from the sum of all possible arrangements of $(n-1)$ elements. The number of arrangements containing false information is merely N minus this number of arrangements possible from $(n-1)$ elements. Similarly, the number of arrangements containing false information as a result of two erroneous quantities is N minus the number of arrangements possible from $(n-2)$ quantities. An example makes this clearer.

> If $n = 4$, and 1 element is incorrect, then the number of arrangements in N that contain erroneous information can be calculated as follows:
>
> Number of arrangements containing error $= N_4 - N_3$
> $= [4 + 4(3) + 4(3)(2) + 4(3)(2)(1)] - [3 + 3(2) + 3(2)(1)]$
> $= 64 - 15$
> $= 49$ arrangements containing error

The following table contains a few numbers that indicate the advantages of correct information to the problem-solver. The first column represents the number of elements available to combine as Mr. Koestler would like us to. The second column indicates the number of arrangements available from the *n* elements. The third column gives the number of arrangements which contain erroneous information if one of the elements (*a*) contains error. The fourth column gives this number if two of the elements (*a* and *b*) contain error.

n	Possible arrangements	Erroneous if *a* is wrong	Erroneous if *a* and *b* are wrong
1	1	1	
2	4	3	4
3	15	11	14
4	64	49	60
5	325	261	310
6	1,956	1,631	1,892
7	13,699	11,743	13,374
8	109,600	95,901	107,644
9	986,409	876,809	972,710
10	9,864,100	8,877,691	9,754,500

These simple-minded numbers are not intended to be a model of conceptualization. I merely throw them in to demonstrate how rapidly combinations containing erroneous information build up as incorrect elements are introduced.

During the solution of a problem correct and adequate information is, of course, extremely important. An intellectual block that may prevent the problem-solver from acquiring well-balanced and pertinent information can be disastrous. Mechanical engineers with a block against electrical engineering or electrical engineers with a block against mechanical engineering may design strange things, such as mechanical television sets or complex electrical power transmission systems where simple mechanical ones would be cheaper and more reliable. People who consistently resist utilizing mathematics limit their problem-solving abilities by being blocked from useful quantitative data. Just as people who are blocked against considering aesthetic, emotional, and qualitative inputs in their decision-making also limit their problem-solving capabilities by refusing to acquire often useful information. Engineers who are uncomfortable with aesthetics can make outstandingly inhumane and ugly devices which may, as a side issue, not even sell well. Environmentalists who ignore the use of quantitative facts and statistics cannot be very productive in designing effective solutions to environmental problems.

There is, however, disagreement as to whether information is universally valuable at all phases of problem-solving. One school of thought maintains that one of the worst enemies of innovation is the large impact of existing solutions on conceptual thinking. This is the school that says: "It is difficult to think of alternate methods of felling small trees if you have spent a lot of time swinging an axe." I know one extremely inventive engineer who finds it very important to operate with a "clean" mind—he avoids learning anything about previous, related solutions to his problems. However, I know another equally productive engineer who spends a great deal of effort learning everything he can about every previous development that seems even slightly related to his problem (a "dirty" mind?). It is true that if you do not know about axes, your solution to felling small trees may be reinventing the axe. You are also denied the use of the axe as a source of additional concepts.

In my opinion, the optimal situation in problem-solving is to be able to use a clean-minded approach to a problem, even though your mind is stuffed with information. I am, of course, biased by my own preferences. As I previously admitted, I grunt my way through problems instead of solving them in an effortless flash of insight. The more information I have about the problem and previous attempts to solve it, the better I do. However, it is sometimes necessary in the problem-solving process to hold this information at arm's length. Certainly, for instance, a massive amount of information is necessary when working with high technology, complex business situations, or interpersonal interactions. However, this abundant information can often prevent you from seeing very elegant solutions. Information makes you an expert, and William J.J. Gordon in *Synectics* says this about expertise: "The specialized semantics of established knowledge constitutes conventions which make reality abstract and secondhand. Learned conventions can be windowless fortresses which exclude viewing the world in new ways."

I believe that it is possible to be an expert and still view the world in new ways. One does not need someone who grew up alone on a desert island to invent a better can opener. One can use people who not only are quite knowledgeable about electrical, mechanical, physical, chemical, and whatever other phenomena, but who also have been closely associated with presently existing can openers. It is only necessary that these people be able to view the world in new ways in spite of all of their prior knowledge. If they can do this they should do better than someone from a desert island.

Expressive Blocks

Turning now to expressive blocks, let us begin by doing another simple exercise.

Exercise: This will require you to find (or make) a simple object whose shape cannot be described by a common name. It could be a block with a corner cut off and a groove along one face, a part from a machine, or any other object with a simple yet irregular three-dimensional shape. Do not use a pencil, a pair of scissors, a tonic water bottle, or other object which is so utilitarian and well known that its shape is familiar to everyone. Find several people, place the object you have chosen in a large paper bag, and have one of the people place his hand in the bag, without looking at the object. He is to describe the object to the other people, who are to draw it.

This exercise is surprisingly difficult. The lack of feedback in the communication loop is of course a contributor to this. Some feedback can be obtained by allowing those drawing the object to ask questions of the describer, although the exercise is most impressive when questions are not allowed. The exercise is also difficult because it is not easy to identify shapes by feel. However, the chief difficulty is probably that of describing a physical object verbally. If your volunteers are mathematically oriented and communicate in terms of x-y-z-coordinates or other geometrical surface description techniques, they will do better at this task. However, if the common verbal approach is used (e.g., "the bottom is a rectangular place with the corner cut off, and then there is a short side going up from the cut-off corner") the task is abysmally difficult. Another reason for the difficulty of the exercise is the rather low level of drawing talent developed in most people. Even if the shape could be perfectly described verbally, most people do not have the faculty to capture it on paper.

I usually do this exercise with a large group so that I can compare drawings after it is over (try it at a party). The presence of the audience adds some interesting emotional blocks. The person describing the object will do better if he or she spends some time feeling the object before describing it. However, it is difficult to spend this time in front of an impatient crowd. The person usually will plunge right into describing. The description will usually proceed at a rapid rate (even though the audience may be picking up no information of use) as the describer usually feels somewhat embarrassed standing in front of a group with his hand in a bag doing what to him may seem a trivial task. Since he will think he has a good idea of the shape of the object (he has his hand on it), he will find it difficult to believe that the audience does not. He may become impatient. He will undoubtedly demonstrate a form of incubation by thinking of a better way to have done his task after the exercise is over.

This exercise demonstrates both the use of *inadequate language skill* to express an idea and the *imprecision in our verbal expression.* It is an extremely common block that one finds often, for instance, in the engineering profession. Many students and engineers are not fond of drawing, partly because they may find it difficult and partly because in some fields drafting has somehow been given a lower status than, say, analysis. We find continual attempts, therefore, to communicate geometrical ideas verbally. Often the degree of difficulty induced by this expressive block is not even appreciated, since the describer knows exactly what he is trying to describe, and the describee often naturally assumes that he understands exactly what the other man is describing. Another problem that demonstrates this block of imprecise verbal expression (if you have access to a dozen people or so) is the following.

> **Exercise:** Give a person a drawing of a simple object (once again an abstract object so a name does not describe its shape). Ask him to look at it awhile and then describe it verbally to another person. The second person should then describe it verbally to a third person, and so on. This should be done in a manner so that the others in the group cannot overhear the descriptions. When the description of the object has been passed through ten people or so, have one last person draw the object. Comparison of the final drawing with the original drawing should prove fascinating.

Other examples of expressive blocks are easy to find: the frustration of trying to present concepts in a foreign language over which one has poor control is typical; the frustration of the writer who is an accomplished typist when his typewriter is broken and he must revert to longhand; the frustration of the executive whose Dictaphone is away for repairs or whose favorite stenographer is sick; the frustration of the non-computer user in today's society. These are all situations in which slowness of expression impedes the problem-solver.

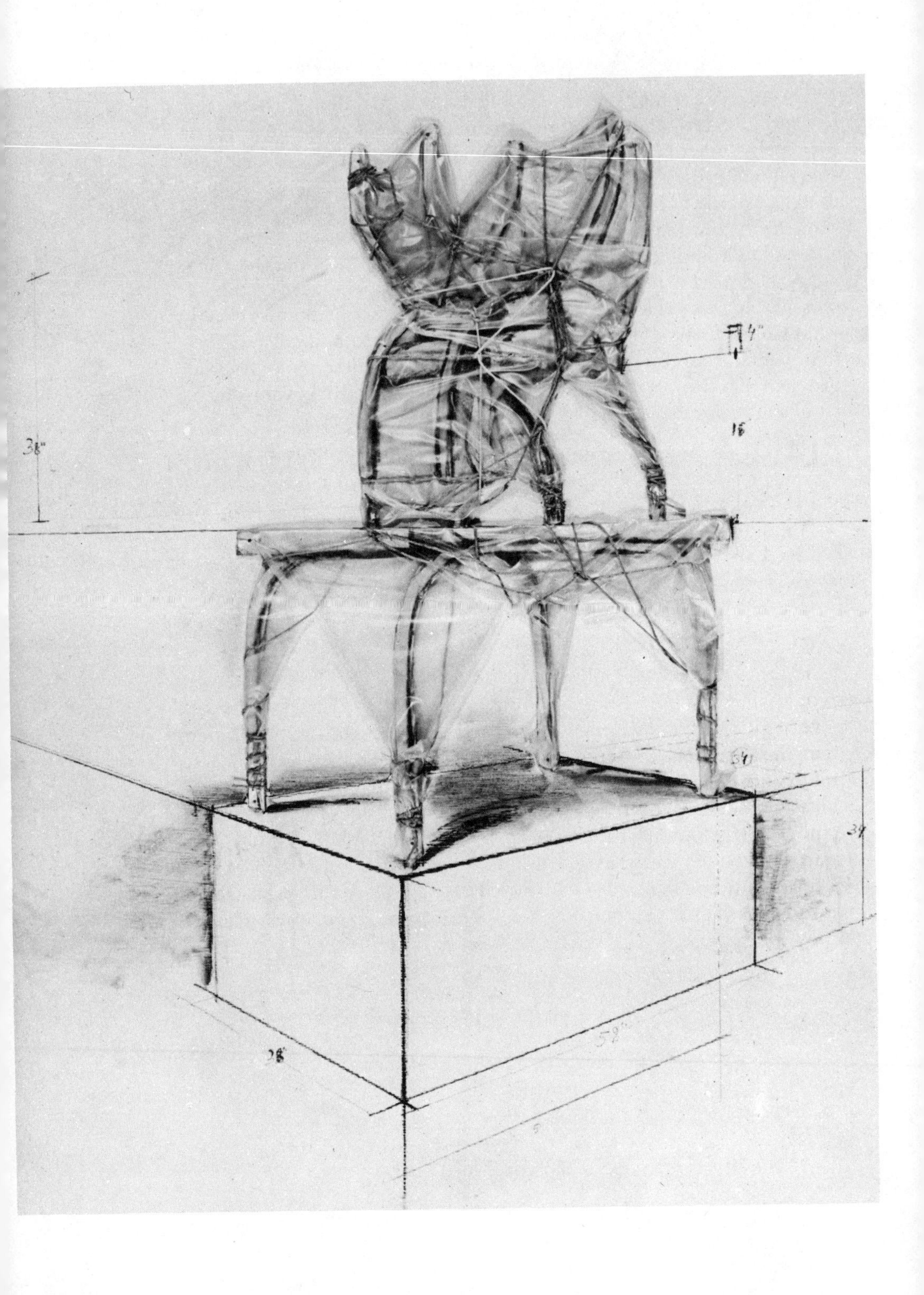
38"
4"
16
28"

CHAPTER SIX

ALTERNATE THINKING LANGUAGES

IN CHAPTER FIVE we discussed the conceptual block resulting from the improper choice of a problem-solving "language." In this chapter, I would like to elaborate on this point. The well-armed problem-finder/solver is fluent in many mental languages and is able to use them interchangeably to record information, communicate with the unconscious, and consciously manipulate. Some of these modes are more "natural" to us than others. They are often even more powerful when used in combination with each other than when used alone.

In this chapter I will discuss some of these thinking modes or languages and put in a plug for some that I do not think receive their fair share of emphasis. To introduce this discussion, let me give you the following exercise:

> **Exercise:** Imagine that you just gave a ride to a hitchhiker who turned out to be an eccentric wealthy builder. As a token of his gratitude, he offers to build an addition onto your house according to your specifications, asking only that the total budget not exceed $12,000. Conceptualize the addition you would ask him to build for you. As you work on this problem try to observe what is going on in your mind (concerning the addition, not the probability of the situation occurring).

Once again, you should have become aware of the difficulty in observing your thinking process as it swings back and forth between the conscious and the unconscious. However, were you roughly aware of what "languages" you employed? Did you think verbally? Quantitatively? Pictorially? Did you imagine smells? Sounds? Tactile sensations? Muscle sensations? Did you tend to work mostly in one language?

If you are typical of most people, you will most easily recall the thinking you did that was in a verbal mode. Verbal thinking is the most prestigious (and perhaps most common) mental language in our culture.

Many psychologists and general semanticists feel that verbal languages are the basis of thinking. For instance, L.S. Vygotsky, in *Thought and Language*, says, "Thought is born through words." Edward Sapir in *Language* says that "language and our thought grooves are inextricably interwoven, are, in a sense, one and the same." Our educational systems reinforce this bias. As Rudolf Arnheim says in his essay "Visual Thinking," in *Education of Vision* (edited by G. Kepes): "In our schools, reading, writing, and arithmetic are practiced as skills that detach the child from sensory (as opposed to verbal or mathematical) experience. . . . Only in kindergarten and first grade is education based on the cooperation of all the essential powers of the mind. Thereafter this natural and sensible procedure is dismissed as an obstacle to training in the proper kind of abstraction." In our culture, we find much emphasis on reading speed and comprehension, on I.Q. tests which rely heavily on verbal ability, and on the use of verbal aptitude scores as an extremely important indicator of intelligence relating to academic and professional potential.

Being a verbal person, I would be one of the last to impugn the sagacity of those who would sanctify the word. It is certainly true that many problems can be well solved verbally. Such solutions can then be easily communicated through well-established verbal channels. However, as we saw in our monk puzzle and our paper-folding problem, there are also problems that can be solved verbally only with great difficulty. With this in mind consider the following problems.

> **Problem One:** Bob has three times as many pine cones as Dan. Between them they have 28 pine cones. How many does each have?

This problem can be worked verbally with the aid of logic, using a trial and error approach between the alternative possibilities. The solution to the problem is relatively simple. It can be solved algebraically as follows:

let b stand for the number of Bob's pine cones
let d stand for the number of Dan's pine cones

We know that the following relationships hold:

$$(1)\ b + d = 28$$
$$(2)\ b \qquad = 3d$$

Plugging in the second into the first and collecting terms, we find that $4d = 28$. Dan must therefore have 7 pine cones and Bob 21.

For any of you familiar with algebra, the mathematics involved here is trivial. Most of you, algebraic or not, are probably able to solve problems like this without too much distress. However, consider the next problem.

> **Problem Two:** Mary has three times as many pine cones as the pine cones owned by both Nora and Oscar. Dan has two times as many pine cones as Bob. Mary has one-and-a-half times as many pine cones as Dan. Oscar and Dan together have as many pine cones as the number Nora has plus twice the number Bob has. Bob, Dan, Mary, Nora, and Oscar have 28 pine cones between them. How many does each have?

I warn that if you are not familiar with algebra, this one will cause you a bit more distress. With sufficient trial and error work, it can be solved through logical verbal thought. However, the amount of work and bookkeeping is annoying, considering the unprofound nature of the problem. A mathematical approach is clearly advantageous. It can be solved algebraically as follows.

As we did before, let the number of Mary's pine cones be "*m*," the number of Nora's be "*n*," and the number of Oscar's "*o*." the relationships between these numbers (including Dan's "*d*" and Bob's "*b*") are:

$$(1)\ m = 3(n + o)$$
$$(2)\ d = 2b$$
$$(3)\ m = 3/2\ d$$
$$(4)\ o + d = n + 2b$$
$$(5)\ b + d + m + n + o = 28$$

As the mathematicians among you realize, these relationships give us enough equations to allow us to solve for all of the unknown numbers. A way (not the only way) to solve these is to first of all substitute the value for $n + o$ shown in relationship (1) into relationship (5), [i.e., $n + o = m/3$]. This gives:

$$b + d + 4/3\ m = 28$$

From relationship (2), we can see that $b = d/2$. Substituting this into the above, we get:

$$3/2\ d + 4/3\ m = 28$$

Relationship (3), however, tells us that $m = 3/2\ d$. Substituting this into our expression, we now get:

$$7/2\ d = 28$$
$$d = 8$$

Hence, Dan has 8 pine cones. Knowing this, we can then happily return to our original relationships and find that Bob has 4 pine cones [from relationship (2)], Mary has 12 [from (3)], and Nora and Oscar together have 4 [from relationship (5)]. Since relationship (4) now tells us that Oscar and Nora have the same number of pine cones, they have 2 each.

Some very simple mathematical language therefore leads us to an answer that requires a good bit of pain to reach verbally. Now you're ready for our third problem.

> **Problem Three:** Suppose that Nora, in a fit of pique, finds a ladder with which she plans to acquire as many pine cones as Mary. The ladder is 10 feet long and the tree trunk is vertical. Just as Nora reaches the top step of the ladder, it begins slipping. Suppose we wanted to know how fast Nora is dropping when the base of the ladder is 6 feet from the tree and is skidding along the ground at 5 feet per second, as shown below.

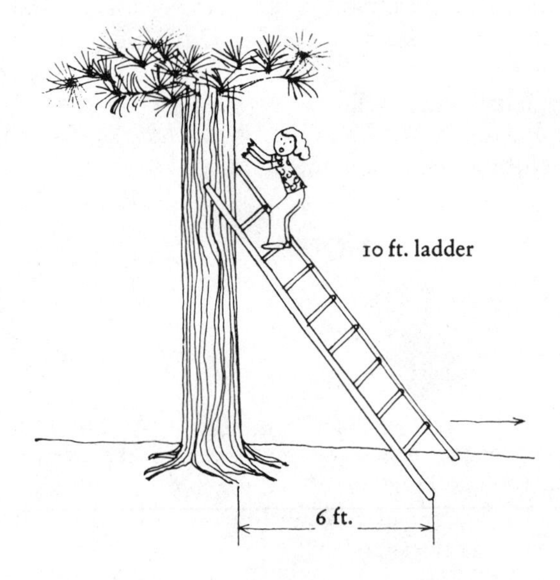

How would you do on this one verbally? Probably not well at all because there is very little to guide you, and a trial and error approach is difficult. Once again, mathematically it is simple, requiring only a trace of trigonometry and calculus. If you abstract the situation as follows:

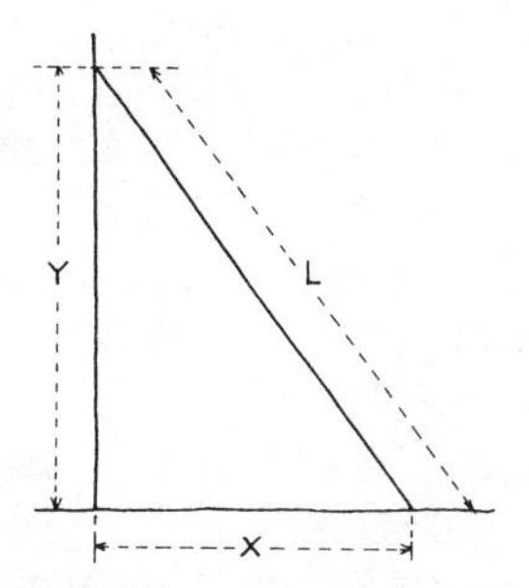

and call the distance from the base of the tree to the base of the ladder x and the distance from the ground to the top of the ladder y, you can apply the theorem of Mr. Pythagoras to write the relationships between these two quantities and the ladder length as:

$$x^2 + y^2 = l^2$$

Substituting 6 feet for x and 10 feet for l, we find that y is 8 feet. Now if we differentiate (a basic calculus operation) this relationship with respect to time, we get:

$$2x\dot{x} + 2y\dot{y} = 0$$

where the dotted quantities are velocities. Substituting 6 feet for x, 8 feet for y, and 5 feet per second for $\dot{x}$, we find that $\dot{y}$ (the quantity we want) is 3¾ feet per second. Once again, simple in the mathematical language—difficult in the verbal.

Clearly, if you attempt to predict the behavior of objects in space, components of complicated machines or structures, or populations and resources, you must include mathematics among your thinking modes. If you cook from recipes, balance your bank account, or use the directions on your lawn fertilizer, you must also use mathematics. If you do not use mathematics as a thinking mode, you are handicapped in working with problems that demand quantification.

Although mathematical aptitude and performance are highly respected in certain circles in our society, verbal aptitude and performance are more generally admired. In fact, in some circles, mathematical illiteracy seems to be a desirable characteristic. Some people seem to feel the cultural need to reject mathematics and boast about their ineptness with quantitative matters. Some eschew mathematics as though it were automatic and without soul, a misconception, of course, since pure mathematicians are motivated and guided by a highly developed sense of aesthetics. Still, mathematicians are heavily stereotyped in the U.S., and mathematical fluency is much less important than verbal glibness in the majority of high-reward positions in this country. For instance, were I to run for the office of President of the U.S. (assuming that this is a high-reward position) I would probably not challenge my opponent to a mathematical problem-solving contest on national TV. In fact, if anything, I would probably conceal my mathematical ability in order not to lose the votes of all those who rejected math as disagreeable in their childhood.

If more people would utilize mathematics in problem-solving (even at a low level of competence) the overall quality of solutions would benefit. Mathematical and verbal thinking together allow much more powerful attacks on problems than verbal thinking alone. I will not further elaborate on the usefulness of mathematical thinking, since it is generally accepted. I discussed verbal and mathematical thinking rather to demonstrate that two thinking "languages" make one a more potent and sophisticated problem-solver, and that some languages are in higher repute (and more frequently relied upon) than others.

I would now like to discuss some languages that we in the Design Division feel are extremely valuable in conceptualization, and are used even less frequently than mathematics. These are the languages of the senses: sight, sound, taste, smell, and touch.

Visual Thinking

A particularly important mode of thinking, which I have referred to several times before and which is presently receiving increased attention academically, is visual thinking. For an excellent treatment of this subject read Bob McKim's *Experiences in Visual Thinking* and Rudolf Arnheim's *Visual Thinking*. Visualization is an important thinking mode which is especially useful in solving problems where shapes, forms, or patterns are concerned. Arnheim explains: "Visual thinking is constantly used by everybody. It directs figures on a chess board and designs global politics on the geographical map. Two dexterous moving men steering a piano along a winding staircase think visually in an intricate sequence of lifting, shifting, and turning . . ." All of us are used to using visual imagery in some situations. For instance, visual imagery is extremely common in dreams. It is also common if someone asks us a question about the appearance of a person or a place. But it is also used in conceptualization, at times when you would not obviously expect its use.

In *The Act of Creation* Koestler quotes Friedrich Kekule, the famous chemist who discovered the structure of the benzene ring in a dream after having devoted a great deal of conscious thought to its enigmatic structure. Kekule describes the discovery:

> I turned my chair to the fire and dozed. Again the atoms were gamboling before my eyes. This time the smaller groups kept modestly in the background. My mental eye, rendered more acute by repeated visions of this kind, could now distinguish larger structures, of manifold conformation; long rows, sometimes more closely fitted together; all twining and twisting in snakelike motion. But look! What was that? One of the snakes had seized hold of its own tail, and the form whirled mockingly before my eyes. As if by a flash of lightning, I awoke.

The result of the dream was Kekule's brilliant insight that organic compounds such as benzene were *closed rings* rather than open structures.

In *Experiences in Visual Thinking* Bob McKim writes of three kinds of visual imagery that are necessary in effective visual thinking. The first, *perceptual imagery*, is sensory experience of the physical world; it is what one sees and records in his brain. The second is *mental imagery*, which is constructed in the mind and utilizes information recorded from perceptual imagery. The third type is *graphic imagery*. This is imagery that is sketched, doodled, drawn, or otherwise put down in a written communicable form, either to aid in your own process of thinking or to aid in communication with others.

Let us first of all briefly consider perceptual imagery, or seeing. By asking you to draw a telephone dial in Chapter Two, I hope I convinced you that you do not record everything you look at, at least at an accessible level (under hypnosis, you might be enabled to draw the dial properly). People see poorly for several reasons. As previously mentioned, one reason is an oversaturation of input. Another is lack of motivation. People tend to see better those things which are more important to them, more unusual, or of an easily-recorded visual character.

You can learn to see better through conscious effort, especially if you are convinced that seeing better is important to you. One way of rapidly developing your *seeing* ability is to engage in activities where you must reproduce things you have seen.

> **Exercise:** You can exercise your seeing ability by looking at things and then drawing them. Such an activity requires not only seeing but imagining and drawing, which will be discussed later. Try this procedure with objects around you, or better yet, objects in your profession that you think it would be helpful to know more about.

A drawing course can improve your seeing ability. If you have to draw trees, you will really start seeing them. I took an art course once in which the teacher took delight in asking us to make quick sketches of friends, family, pets, home, and neighborhood. I found this extremely interesting because I looked at my immediate environment at least two orders of magnitude more closely. One of my colleagues took a photography course in which the instructor taught the students to photograph scenery by taking a jar of beans and the students into a field, throwing a bean into the field for each member of the class, and then telling each student to stand on his bean and spend the day shooting scenic pictures. Such exercises *make* you see. You can take pictures of the Grand Canyon or other scenic wonders without putting a great deal of effort into detailed seeing. However, taking a beautiful picture while standing on your bean in a field requires that you truly use your powers of visual perception.

Now let us talk about the second type of visualization: mental imagery. These mental images are probably the most important for the conceptualizer. According to McKim, there are two aspects of visual imagery which are important. The first he calls clarity (how sharp and filled with detail are the images?) The second he calls control (how well can you manipulate them?). Here is an exercise to let you evaluate your visual "imaging" capability.

> Exercise: *Clarity of Mental Images.* Imagine the following. After each mark a clear (c), vague (v), or nothing (n) in accordance with how clear (sharp and detailed) the image appears in your mind.
>
> 1. The face of a friend
> 2. Your kitchen
> 3. The grille on the front of your car
> 4. A camellia blossom
> 5. A fiddler crab
> 6. A Boeing 747
> 7. A running cow
> 8. The earth from orbit
> 9. Your first car
> 10. Richard Nixon

The clarity of your mental images depends upon several factors. First of all, in this type of exercise, it depends on *seeing*. If you have never seen a fiddler crab or a running cow, your mental images were probably not too sharp. It also depends on your seeing *ability*. And this in turn, as we have mentioned, depends on motivation (your camellia was probably clearer if you are a camellia freak), the visual character of the object (Richard Nixon was probably fairly clear because he has been characterized so often in the news and in political cartoons), timing (your first car may have grown dim by now), and saturation (your car grille?). Finally, it depends upon the image-reproduction mechanism of your brain. There is certainly individual variation in the ability to visually imagine which goes beyond the variability mentioned above. If you ask a roomful of people to visualize a brick or an apple, and then ask individual members of the room questions about their image, you will get a range of answers, the clarity of which extends from images vivid in color, detail, texture, background, and shadows, to no particular image at all.

> **Exercise:** For your own information try visualizing a series of objects and see if you can determine a pattern in your own imaging ability. Are you better at visualizing people than objects? Or worse? Are you better at two-dimensional objects than three-dimensional? Are you better at small things than large things? Where do you see your image? Is it out in front of your eyes or back in your skull somewhere?

Visual imaging ability is complex, since it depends not only upon your ability to form images, but also upon the supply of pertinent imagery which is stored in the mind. However, it seems safe to say that you can improve your visual imaging capability by devoting effort to it and making it a higher priority item in problem-solving. Visual images can be consciously enhanced. When I was a student of John Arnold at Stanford, he was constantly hitting me with "visualize an apple" type problems. As a result, I became so conditioned that when asked to visualize something I still concentrate all of the information and energy I can on the task. Now let us look at your ability to *control* (manipulate) visual imagery.

> **Exercise:** Imagine the following:
> 1. A pot of water coming to a boil and boiling over
> 2. Your Boeing 747 being towed from the terminal, taxiing to the runway, waiting for a couple of other planes, and then taking off
> 3. Your running cow changing slowly into a galloping racehorse
> 4. An old person you know well changing back into a teenager
> 5. A speeding car colliding with a giant feather pillow
> 6. The image in (5)—in reverse

Are you better at manipulating images you have actually seen or in creating new ones? Can you modify images in a fantastic (non-real) way? Take some time, and see whether you can extend your understanding of your ability to control your visual imagery. Try manipulating various types of images, inventing images in your mind, etc. Many people feel that the ability to control visual images can be developed through practice. Bob McKim in *Experiences in Visual Thinking* discusses what he calls "directed fantasy" as a way of strengthening imagination. In "directed fantasy" the participant is asked to fantasize in a number of directions that take him through a wider range of imaginative activities. He is forced to "exercise" his imaginative abilities and confront imaginative blocks that he would ordinarily avoid. By finding that he is able to wander freely through these areas and allow his imagination to range widely without catastrophic results, he becomes encouraged to feel more familiar with the use of visual imagery in conceptualization.

Now let us discuss the third type of visualization—graphic imagery. In order to take full advantage of visual thinking ability, *drawing* is necessary. Drawing allows the recording, storage, manipulation, and communication of images to augment the pictures you can generate in your imagination. In the Design Division, we find it useful to divide drawing into two categories: that which is done to communicate with others, and that which is done to communicate with oneself. The following drawing is a type used generally for the purpose of communicating with others. It was drawn by architect Walter Thomason of San Francisco.

Drawings such as these idea sketches are generally used in communicating with oneself. They were drawn by a Stanford engineering graduate student, Peter Dreissigacker. (Yours need not be as artistic.)

The first type of drawing (communicative) receives a good bit of attention educationally, and you can learn to make such drawings through formal courses of instruction. The second type (thinking sketches) receives far less emphasis, yet it is an important adjunct to visual thinking. Given a large pad of paper and a pencil, most people will make sketches as they work on sample problem exercises. Strangely enough, the same people often will not go to the trouble of summoning their drawing materials on their own in a problem-solving situation.

I am probably sensitized to these particular problems because I have spent quite a bit of time teaching design in a university setting which attracts an extremely verbal group of students. A great deal of effort has been put into their verbal (and mathematical) abilities during their formal education, but little into their visual ability. When they come to Stanford many are "visual illiterates." They often are not used to drawing, nor to using visual imagery as a thinking mode. Although their drawing is generally not good, it is usually good enough (especially with a few helpful hints) to use as a thinking aid. Nonetheless, they are usually extremely reluctant to draw because their drawings compare so badly with drawings made by professionals (intended for communication with others). In design, we try to encourage crude but informative drawings for the student's own purposes. We also try to encourage improving one's drawing skills, since we find that good drawing skill is a powerful conceptual aid. Try the following exercise and see whether your drawing skills (no matter how marginal) help you in conceptualizing.

> **Exercise:** Buy a cheap notebook of a convenient size (small enough to accompany you, but as big as possible otherwise) and provide yourself with the most satisfying line-maker you can find. (A good one is a Pentel-type pen, which looks like a ballpoint pen but has a tapered fiber tip that makes an instant dark and smooth line.) Make drawings for yourself in this notebook for the next week or so while you are conceptualizing and otherwise involved in solving problems. Your drawings may be doodles, block diagrams, schematics, squiggles, sketches, or what have you. Try to see which of these drawings (if any) help in problem-solving and which do not. Are they of more use in particular portions (e.g., at the beginning) of the problem-solving process? Does a lack of drawing-skill minimize their effect? Do you use your notebook to refer to your previous work? Does the size of your notebook inhibit you? (If so change to newsprint or butcher paper on the wall or a table and to larger felt pens or to color, and try again.)

By emphasizing drawing as a thinking technique, I do not mean to belittle its power as a communication device. I have had many experiences in which people who are able to draw well have been able to influence others in a problem-solving situation, for better or worse. This is especially prevalent in design situations where no precedent exists. I was recently supervising a student group engaged in the design of a new type of underwater vehicle. One of the students was excellent at making quick renderings. Each time he would put forth a concept it would seem so real in its rendered form that the group would gleefully adopt it. Then, when he produced another concept the next day, the group would be in temporary consternation until they adopted the new one. I have seen the same thing happen many times in the design of spacecraft, where little visual precedent exists. A well-drawn concept has amazing power. Drawings, of course, also have amazing ability to convey precise information, even if crudely done. One of my oldest and best friends, for instance, is a farmer. When I visit him I often accompany him on his rounds. Since I am a helpful sort, I try to aid him in his various projects, which often consist of moving large inert objects. Such tasks should be straightforward to someone like me, who is, after all, an engineering professor. However, he gives directions verbally, with the result that I am always bewildered by a version of the situation we encountered in the block-in-the-bag exercise in Chapter Three. (I am on a long-term project of getting him to draw crude pictures for his people and for me, so I will not stupidly put things in the wrong place.)

Exercise: Use your drawing skills, no matter how marginal, to aid you in giving directions to people. Carry a small pad and pencil to do this if necessary. You are probably familiar with drawing crude maps to show people how to get to your house. But have you ever drawn your children a map of where to pile the leaves they are raking up, your husband a map of how to put things away in the house, or your wife a map of how to carve a roast (sorry for any role typing)?

Let us now leave visual imagery, an extremely important tool in conceptualization. If you have the time and inclination to attempt to develop your visual thinking ability, the Reader's Guide will give you places to start. At least become aware of your abilities and limitations with visual imagery and attempt to use visualization in your thinking process whenever appropriate. It is one of the most basic of all thinking modes and one that is *invaluable* in problem-solving.

Other Sensory Languages

We will now go on to other sensory languages that are essential in conceptualization and are used even less frequently in general problem-solving than visual thinking. Just as visual imagery corresponds to the sense of sight, other types of sensory imagery also result from their corresponding senses.

Here are some exercises for you to test how good you are at different types of sensory imagery. Once again, rate them clear (c), vague (v), or nothing (n).

Exercise: Imagine:
The laugh of a friend
The sound of thunder
The sound of a horse walking on a road
The sound of a racing car
The feel of wet grass
The feel of your wife's/husband's/girlfriend's/boyfriend's/pet's hair
The feel of diving into a cold swimming pool
The feel of a runny nose
The smell of bread toasting
The smell of fish
The smell of gasoline
The smell of leaves burning
The taste of a pineapple
The taste of Tabasco sauce
The taste of toothpaste
The muscular sensation of pulling on a rope
The muscular sensation of throwing a rock
The muscular sensation of running
The muscular sensation of squatting
The sensation of being uncomfortably cold
The sensation of having eaten too much
The sensation of extreme happiness
The sensation of a long attack of hiccups

Now try the following for control of separate sensory images:
The sensation of being uncomfortably cold changing to one of being uncomfortably hot
The laugh of a friend changing into the sound of thunder
The feel of wet grass changing into the feel of your wife's/husband's/girlfriend's/boyfriend's/pet's hair
The smell of fish changing into the smell of gasoline
The muscular sensation of pulling on a rope changing into the muscular sensation of rowing a boat

Such exercises have a function equivalent to the earlier exercises on visual imagery. They may help you develop your sensory imagery ability, if used extensively. In any case, they at least let you learn more about your ability to image in various sensory languages.

Sight tends to be the predominant sense from a physiological standpoint. However, just as verbal thinking should not be allowed to elbow visual thinking out of the way, neither should the visual mode be allowed to overpower other sensory modes. Smell, sound, taste, and touch are extremely important to problem-solvers for three reasons:

1. Since they are low on the thinking "prestige" list in our culture they can lead you to innovative and overlooked solutions. (Tarzan had a well-developed sense of smell, but I am sure that no one would *expect* the same from a Nobel Prize winner.)
2. They are necessary for the solution of problems in which smell, sound, taste, and touch are involved (the design of a new hors d'oeuvre).
3. They augment visual imagery and each other to vastly increase the clarity of one's total imagery (more about this later).

Let us briefly discuss the first reason listed above. I often give my students problems having to do with developing devices to help blind people. I do this both because it proves highly motivating and requires ingenuity, and because it makes them think about various types of sensory inputs. Most of them attack the problem initially by imagining that they are blind. This is difficult to do, because sight is such an overwhelming input to most people that they find it hard to take the role of a blind person using their imagination alone. They especially find it hard if they generally think only verbally or mathematically.

After letting them work on the problem a while, I blindfold the students for an hour or two and let them wander about the world. This gives them, all at once, a chance to accept input from their other senses. They are then much more likely to use this very important data (to a blind person) to solve their problems. This simulation of blindness has limited accuracy, because when you are blindfolded for an hour or two your main problems have to do with walking, whereas blind people have long since overcome this. Still, the simulation is effective in bringing the awareness of messages from the other senses to people.

Exercise: Try this experience yourself. Find someone to keep you out of trouble (not to physically guide you, just to keep you off the freeway, out of open manholes, poison oak, etc.), blindfold yourself, and walk around for an hour or so. You will be amazed at the sensory data you accumulate.

The second reason for using all the senses, "They are necessary for the solution of problems in which smell, sound, taste, and touch are involved," should be obvious. Just as an architect is better off with a good ability to image spaces and forms, so a cook is better off being able to image taste and smell.

The third point, "They augment visual imagery and each other," is more subtle. I am sure that most of you are aware that the senses augment each other. Food is a combination of taste, smell, and sight. Vichyssoise is as unsuccessful when one has a cold as an omelette is if it is dyed blue. An electrical storm needs sound as well as sight to be really dramatic. Sexual excitement benefits from feel, smell, and taste as well as sight and sound. Similarly, mental images need the full dimensionality of all the senses to be most effective.

In order to demonstrate this, I would like to give you another exercise. However, before you begin, I would like you to think of an apple. Got it? Now ask yourself enough questions about this apple that you can establish the clarity of your image.

I am now going to give you the transcription of a tape that Bob McKim and Bill Verplank use in a Stanford class in Visual Thinking. The tape is concerned with *clarity* and *control* of combined sensory imagery. It is played to small groups of students in an isolated and comfortable environment. They are typically sprawled comfortably on a soft, carpeted floor with no distractions present. They have been prepared for the tape by a talk explaining the serious purpose of the exercise and several weeks of class stressing the importance of sensory imagery.

Exercise:

1. Find a sympathetic narrator and preferably a few other people who are interested in this subject.
2. Obtain an outstanding apple for the narrator to give to each person.
3. Relax in a comfortable spot with the others.
4. Have the narrator read the following to you. He should read slowly, seriously, and soothingly, giving you plenty of time to fully establish the images before he goes on. You can help by giving a prearranged signal (a lifted finger?) when you have the image fully developed, at which point he can go on to the next statement.

"First close your eyes and relax. Direct your attention inward. Now imagine your self in a familiar setting in which you would enjoy eating an apple. Relaxedly attend the sensory mood and detail of this place. Now, imagine that in your hand you have a delicious, crisp apple. Feel the apple's coolness, its weight, its firmness, its round volume, its waxy smoothness. Explore its stem. Visually examine details, see bruises, the way sunlight sparkles on the facets of the apple's form, the way the skin reflects a pattern of streaks and dots, many colors, not just one. Attend this image till your mouth waters. Now bite the apple. Hear its juicy snap, savor its texture, its flavor. Smell the apple's sweet fragrance. With a knife, slice the apple to see what's inside. As you continue to explore the apple in detail, return occasionally to the larger context, see your hand, feel the soft breeze, be aware of the three dimensionality of form and space.

"How was your apple this time? Probably a lot better. But your apple is probably still not as vivid as possible, simply because you don't really know what an apple is like. We've all eaten plenty of apples, but how often do we really pay attention. We are most often doing something else while eating, talking, reading, thinking, but never attending to every sensory detail. We're going to give each of you an apple now [hand out the apples] and you can eat it. We're going to ask that no one talk. All of your attention should be on the apple and on your sensations. Before you eat your apple, take a minute to examine it. Look at its shape, its volume, its color, its markings. Feel its temperature, its texture, its firmness, its mass. When you really know it, take a bite from it. Listen, smell, taste, feel, attend every sensory detail. Take your time.

"The apple that you have just eaten is now being assimilated by your digestive system. The apple is becoming you. Imagine that you are the apple that you have just eaten. Imagine that you are an apple on an apple tree. Take a deep breath. Let it out, and as you let it out, relax all tensions. Quiet all distracting thinking. Direct all of your attention, in a very relaxed way, to the pleasurable thought of being an apple on a real apple tree in a beautiful apple orchard way out in the country. You can feel the warm sun on your skin. You can feel a soft breeze. The sky is clear blue. The sun feels good as it radiates into your apple body. You can hear the leaves of your tree rustling in the breeze. You can smell the fragrance of the ripening apple orchard. It feels good to be part of nature. Now imagine that you are regressing in time. You are an apple that is going backward in time, becoming smaller, smaller, greener, tarter, smaller yet, you are evolving in reverse into an apple blossom. You are an apple blossom together with many other apple blossoms on your apple tree. You can smell the lush fragrance of apple blossoms. You can feel the warm sun on your delicate petals. You can hear the honey bees buzzing as they go about pollinating the orchard. In the distance you can hear a farmer's dog barking. You can taste your own sweet nectar. You can feel that you are an integral part of an incredibly complex natural process involving sun, earth, air, bees, the seasons. It feels good. Now you are becoming aware that you are more than a single apple blossom. You are an apple tree. Allow your imagination to move into thc branch that supports the blossom. You can feel the sap that brings energy to the tree's leaves and blossoms. You can feel the sap moving through you. Follow this flow of energy down into the trunk of the apple tree. Feel the strength of the trunk in your own body. You must be strong to support branches loaded with ripe apples, and to resist the force of heavy wind. Feel the rough texture of your bark, the hardness of your wood. Now direct your attention down the trunk and into the roots of your apple tree. Reach out into the dark, damp soil. See the darkness. Smell the fragrance of the fertile soil. See the fat worms and the other subterranean creatures that work the earth. Feel the cool wetness and texture of the moist dirt and rocks, as your roots reach out for life-giving water and nutrients. Now leave the tree. Become the water itself in the damp orchard field. Feel yourself feeding the grasses and the wild flowers. You are part of a larger concept. You are essential to life. You are part of the much larger unity of nature. As water saturated in the orchard field, experience the sun's heat drawing you upward. Feel the sun evaporating your body, transforming your liquid nature into vaporous water. Feel your molecules rising upward into the blue sky toward the blazing sun. You and the others are now forming into a soft cloud. Down below you can see the earth, the tiny patch of the apple orchard, you are floating in the blue sky effortlessly. Quiet, billowy, incredibly free. In the distance a hawk is soaring. You are part of the creative cycles of nature. Now the sky is darkening, becoming cooler, you can feel the wind swirling and moving through your cloud. You are condensing with other molecules into droplets of rain. Falling downward, through the cold gray sky, downward, downward. You splash the leaves of a green apple tree and fall down to the ground, to the soil, to the roots, to the strong trunk, to the sap that feeds the branches, the leaves, the blossoms, the apple. You are the apple on the tree, in the orchard, on a rainy day. You can hear the rain splattering on the leaves, feel the cold stormy wind swaying the tree's branches, smell the rich odor of damp earth. Your apple, created by this marvelous interwoven working of nature, is inside you becoming you. And you, in turn, are a unique part of this creative unity. As you return now to your aliveness, here, and now, you feel good to be part of a unity which is inherently and eternally creative."

The students in the class do not use this tape until they have had quite a bit of experience in conceptualization, visualization, and imaging exercises of various sorts. However, we find that most people without

THE FLIGHT OF INTELLECT

Portrait of MR GOLIGHTLY,
experimenting on Mess. Quick & Speed's new patent, high pressure,
STEAM RIDING ROCKET.

Pub. by C. Tilt, Fleet St.

this background are still able to build a crunchy luscious apple image in the first portion of the tape and enjoy their real apple with heightened sensory awareness. We find that people have more difficulty with the "control" portion of the tape and find it less relevant than the students do. However, it is useful as another indication of your ability to control imagery. You may now get a real apple from the refrigerator (if you haven't already) and eat it as you proceed to the next section, which concerns freeing the subconscious.

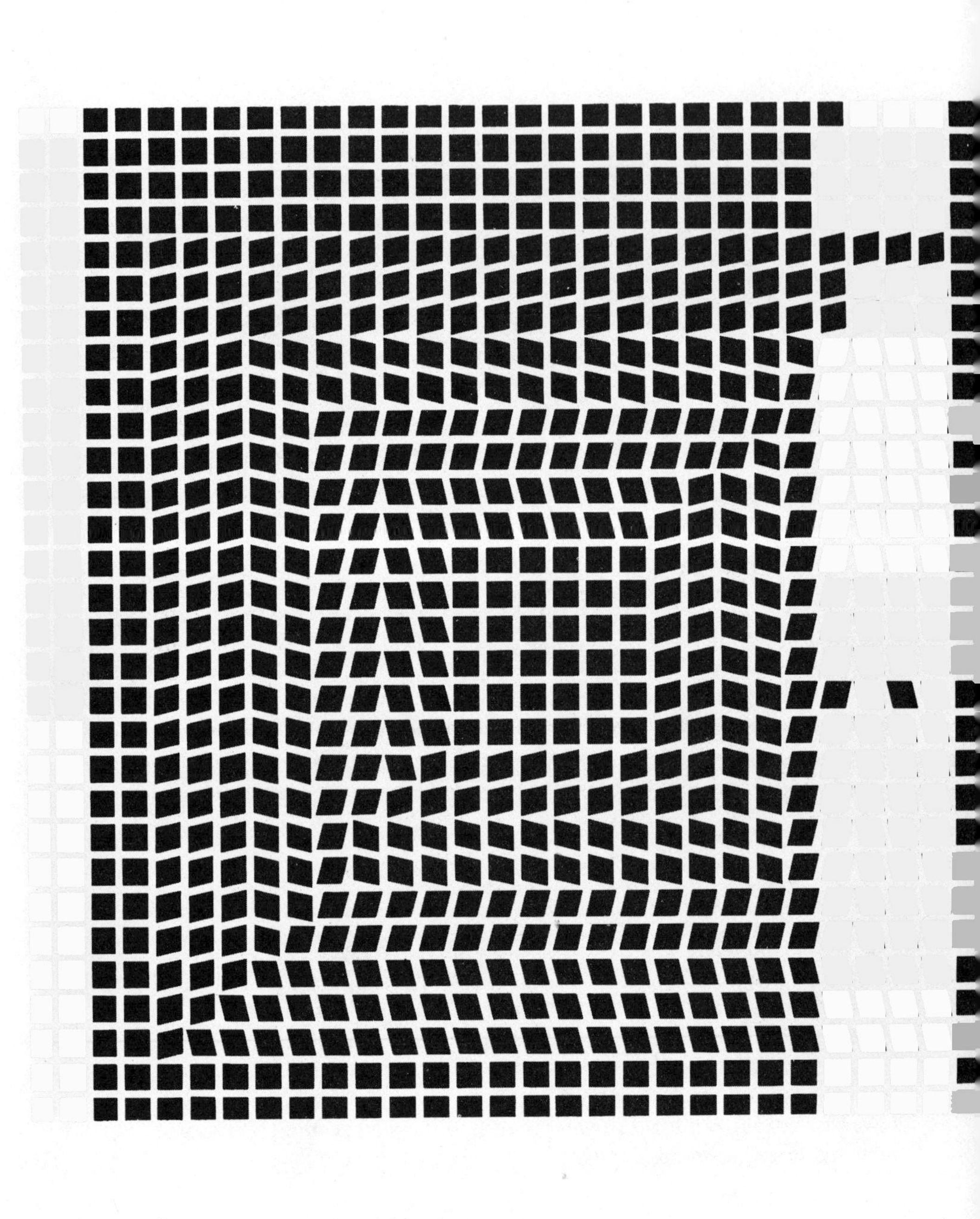

CHAPTER SEVEN

ALL KINDS OF BLOCKBUSTERS

THE USE OF A RICH vocabulary of thinking languages, as discussed in Chapter Six, is one way to overcome conceptual blocks. However, there are many other methods. We will look at a variety of such techniques in this chapter. First we will explore a few techniques that allow the use of the conscious mind to overpower conceptual blocks. In a sense these techniques force thoughts that would not otherwise occur. The last part of the chapter will be concerned with methods to become more relaxed, less critical, and more intellectually playful during problem-solving.

The process of consciously *identifying* conceptual blocks takes one quite a distance toward *overpowering* them. Still, there are specific methods of going further. Many of these blocks exist because of the achievement-oriented, competitive, and compulsive nature of Western man. However, this very combination of characteristics outfits him optimally for consciously overpowering such blocks. We think that people who are interested solely in good grades are often not as creative in school as they could be. However, if they are put in a course that is graded on creative output, they become much more creative. Their motivation and mental discipline are sufficient that they quickly figure out ways to become more creative and act accordingly. Let us talk about a few such methods that can be consciously applied to problem-solving.

A Questioning Attitude

One of the most important capabilities in a creative person is a questioning attitude. Everyone has a questioning attitude as a small child, because of the need to assimilate an incredible amount of information in a few years. The knowledge that you acquire between the ages of 0 and 6, for instance, enormously exceeds what has been consciously taught. A great amount of knowledge is gained through observation and questioning. Unfortunately, as we grow older many of us lose our questioning attitude. There are two principal reasons. The first is that we are discouraged from inquiry. After the child reaches a certain age, parents and others are often not as patient with questions (especially if they are busy and/or do not know the answer) that do not seem socially pertinent (Why can you see through glass? Why are leaves green?) and tend to discourage the questioner. Our educational institutions can barely convey the knowledge they are held responsible for (reading, writing, arithmetic, cultural lore). There is little time available for answering questions, so questions are effectively limited and discouraged. Many is the professor who begins his lecture with a plea for questions and then ends it with neither the time nor the encouraging attitude necessary to get them.

The second reason the child's "inquisitive" nature is socialized out of us (or at least diminished) has to do with "the great knowledge game." We learn as we grow older that it is good to be smart. Smartness is often associated with the amount of knowledge we possess. A question is an admission that we do not know or understand something. We therefore leave ourselves open to suspicion that we are not omniscient. Thus, we see the almost incredible ability of students to sit totally confused in a class in a university that costs thousands of dollars a year to attend and not ask questions. Thus, we find people at cocktail parties listening politely to conversations they do not understand, and people in highly technical fields accepting jargon they do not understand. One of my colleagues from my aerospace days used to delight in feeding nonsense jargon and erroneous arguments to people in other specialties. They would seldom question him in sufficient depth to find that he was faking. I have another friend who once successfully delivered a totally fraudulent hour-long lecture in aerospace medicine, of which he is totally ignorant, to an audience of university students. When his true credentials (none) were revealed to the students at the end of the lecture, they immediately voiced doubts that they had accumulated during the hour. They also registered extreme displeasure both toward the speaker and the instructor for violating their trust and wasting their time. However, during the talk itself, the competence and facts of the speaker were not questioned, probably because of the confidence with which he spoke and his extremely articulate lecturing style.

As I previously said, the questioning attitude is necessary in the broadest sense to motivate conceptualization. If you accept the status quo unquestioningly, you will have no reason to innovate. You will not be able to see needs and problems, and problem-sensitivity is one of the more important qualities of the creative person. Once the problem is sensed, the questioning attitude must be used continually to ensure a creative solution. A creative person should have a healthy skepticism about existing answers, techniques, and approaches.

In a fascinating book called *The Universal Traveler. A Soft-Systems Guide to: Creativity, Problem-Solving, and the Process of Design*, the authors Don Koberg and Jim Bagnall discuss what they call "Constructive Discontent":

> Arrival at the age of 16 is usually all that is required for achieving half of this important attribute of creativity. It is unusual to find a "contented" young person; discontent goes with that time of life. To the young, everything needs improvement. . . . As we age, our discontent wanes; we learn from our society that "fault-finders" disturb the status quo of the normal, average "others." Squelch tactics are introduced. It becomes "good" not to "make waves" or "rock the boat" and to "let sleeping dogs lie" and "be seen but not heard." It is "good" to be invisible and enjoy your "autonomy." It is "bad" to be a problem-maker. And so everything is upside-down for creativity and its development. Thus, constructive attitudes are necessary for a dynamic condition; discontent is prerequisite to problem-solving. Combined, they define a primary quality of the creative problem-solver: a constantly developing Constructive Discontent.

This questioning attitude can be achieved by conscious effort. You merely need to start questioning. An emotional block is involved here, since you are apparently laying your ignorance out in the open. However, it is a block that will rapidly disappear once you discover the low degree of omniscience present in the human race. No one has all of the answers and the questioner, instead of appearing stupid, will often show his insight and reveal others to be not as bright as they thought. The most learned man can be overrun merely by continually applying the question "why?" or "how?" Pick a scientist, for instance, and ask him a "naive" question about something in his discipline. A few questions will drive him back through the basic knowledge which exists. Most of the questions you used to wonder about in your youth (What is beyond the farthest star? What is life? Why do people die?) are still unanswered.

In fact, the man who often is most admired at scholarly meetings is the penetrating questioner, who asks the apparently simple question that points out the flaws in a complex theorem or other structure of knowledge. Therefore, you have nothing to lose and a great deal to gain by questioning. The only thing you need to remember is that everyone is not as enlightened about knowledge as you (now) are, and some people will become unhappy if questioned to the degree that their omniscience becomes suspect. ("Why should man be creative?" "Because creativity allows self-actualization." "What good is self-actualization?" "It allows man to be happy." "What is happiness?" "Well-being." "What is well-being?" "Go to hell.")

If you still hesitate to ask questions, here are a few harmless and innocent questions. Ask them of anyone, and you will find that you are not as relatively ignorant as you thought.

1. Why do people sleep?
2. Do mirrors make letters appear backwards? If so—why do they not make them appear upside down?
3. A canary is standing on the bottom of a large sealed bottle that is placed on a scale. He takes off and flies around the inside of the bottle. What happens to the reading of the scale? What if he is a fish and the bottle is full of water?
4. What is licorice made out of? Why is it black?
5. Many cosmologists presently agree that the universe was created by a big "bang," or explosion, and that all of the stars are traveling outward from the original "bang." What preceded the "bang"?

Exercise: Questioning is especially important in problem-finding and problem-definition. You are going to use questioning in this manner. To play the game you need a cooperative person who is in a profession with which you are not very familiar. This exercise may take a reasonable amount of time, but if the person is a friend of yours or is interested in activities such as this, he should not object. Begin by asking him questions and ask them until you have a specific problem in his profession isolated and defined. Don't be satisfied with a vague, overly general, big-picture problem (medical care for the aged is inadequate). Try for a specific problem statement that is obviously soluble with a small amount of effort (for example, the sight of a novocaine needle scares people).

As you ask your questions, be aware of where your difficulties lie. Are certain types of questions (personal?) more difficult to ask than others? Can you observe the difficult period that results when you have used up your "social" questions and have to get down to work? What is your subject's response to different types of questions? Do you find it interesting to find out so rapidly about another profession (you should)? Were you able to go from a very general problem statement to a specific one? Did you work with several problem statements on the way to your final one?

Fluency and Flexibility of Thinking

Fluency and flexibility of thinking can also be attacked consciously. List-making is one of the simplest, most direct methods of increasing your conceptual ability. People often compile lists as *memory* aids (shopping lists, "do" lists). However, lists are less frequently used as *thinking* aids. List-making is surprisingly powerful, as it utilizes the compulsive side of most of us in a way that makes us into extremely productive conceptualizers. It does not require (in fact would suffer from) change in behavior and flourishes in a competitive environment.

Exercise (Part I): In order to give you a better feel for list-making, let me give you an exercise based on the "brick use" test attributed to J.P. Guilford. Imagine that you are a consultant for a brickyard that makes common red construction bricks and is in financial difficulties. The manager of the brickyard is interested in new uses for his products and has asked you to provide him with some. Spend a few minutes (three or four) thinking about the problem and then write down on a sheet of paper a new use for bricks.

Were you aware of what went on in your mind when you were thinking about the problem? You probably did some type of ad-hoc listing of alternatives. However, your conceptualization may have suffered from lack of focus, premature judgment (rejecting ideas that seem impractical), and labeling (choosing only the stereotyped usages).

> **Exercise (Part II):** Now take a blank piece of paper and spend four minutes listing *all* of the uses you can think of for bricks. Remember to aim for fluency and flexibility of thinking and not to get hooked into premature judgment or labeling. Go.

You may have noticed (especially if you did this exercise with others) that people tend to be very intense when listing ideas, particularly when a time limit is involved. This is perhaps a remnant from our educational system and the general competitive nature of our society. However, whatever the cause, listing focuses your conceptual energy in a rather efficient way and produces a written record of the output—both advantageous features. If the above exercise was successful, your "listing" effort should have gotten you much further conceptually than your original non-directed "spend a few minutes" effort. Were you fluent and flexible? As a calibration point in fluency, our design students average between 10 and 20 uses on this exercise. Some produce between 5 and 10, others between 10 and 20, and others between 20 and 30; a few produce under 5 or over 30 the first time through. The curve is roughly bell-shaped. *Fluency*, of course, is not enough in conceptualizing.

If your list were to consist of entries such as "build a wall, build a fireplace, build a patio floor, build a shoestore, build a hardware store, build a clothing store, build a grocery store," and so on, you could have been fluent, but of limited use to the brickyard owner who is probably already familiar with these uses. *Flexibility* of thought is also needed. You are flexible if your list included usages such as the storing of water, the warming of sheets on cold nights, the leveling of dirt, raw material for sculpture, playground blocks for children, and objects for a new track-and-field event (the brick-put). Such usages show an ability to see beyond the conventional role of bricks.

If you were doing this exercise with others, swap lists around and read them. Remember that some of the ideas should strike you as funny, if your flexibility is working well. If your list is lacking in flexibility, you may be suffering from the "premature labeling" block we discussed in Chapter Two. The usage of bricks is heavily stereotyped (construction material).

In Chapter Two we discussed the listing of attributes as a method to escape the inhibiting effects of premature labeling. The listing of attributes is a powerful way to rapidly get more insight into the possible usefulness of an object, which in turn is an advantage in conceptualizing.

Let us list the attributes of a brick. Some of them are:

weight
color
rectangularity (sharp edges, flat faces)
porosity
strength
roughness
the capacity to store and conduct heat
poor capacity to conduct electricity
hardness

Think of more, if you can, and add to the list. How about economic considerations? Aesthetic aspects? I am sure that you can now see that by taking any attribute (weight), it becomes rather easy to list non-conventional uses for a brick (anchor, ballast, doorstop, counterweight, holding down tarpaulins or waste newspapers, as projectiles in wars, riots, neighborhood rumbles, etc.). It is often a help in conceptualization to consider attributes instead of commonly used labels.

Thinking Aids

A clever use of attribute listing is contained in *The Universal Traveler* which authors Koberg and Bagnall call "Morphological Forced Connections." They give the following rules for their "foolproof invention-finding scheme" along with an example showing how their scheme works. Here it is:

Morphological Forced Connections

1. List the attributes of the situation.
2. Below each attribute, place as many alternates as you can think of.
3. When completed, make many random runs through the alternates, picking up a different one from each column and assembling the combinations into entirely new forms of your original subject.

After all, inventions are merely new ways of combining old bits and pieces.

Example.

Subject: Improve a ball-point pen.

Attributes:

Cylindrical	Plastic	Separate Cap	Steel Cartridge, etc.

Alternates:

Faceted	Metal	Attached Cap	No Cartridge
Square	Glass	No Cap	Permanent
Beaded	Wood	Retracts	Paper Cartridge
Sculptured	Paper	Cleaning Cap	Cartridge Made of Ink

Invention: A Cube Pen; one corner writes, leaving six faces for ads, calendars, photos, etc.

Another use of attribute listing, credited to Fritz Zwicky, is called morphological analysis. It is an automatic method of combining parameters into new combinations for the later review of the problem-solver. For instance, in an example done by John Arnold and taken from *Source Book for Creative Thinking* by Parnes and Harding, the problem is to provide a new concept in personal transportation. First of all, three parameters of importance are selected (more than three could be selected, but they could not easily be drawn on a piece of paper and put into this book). Alternate possibilities for the three chosen parameters (motive power source, type of passenger support, and the media in which the vehicle operates) are then listed on three orthogonal axes as shown on the next page.

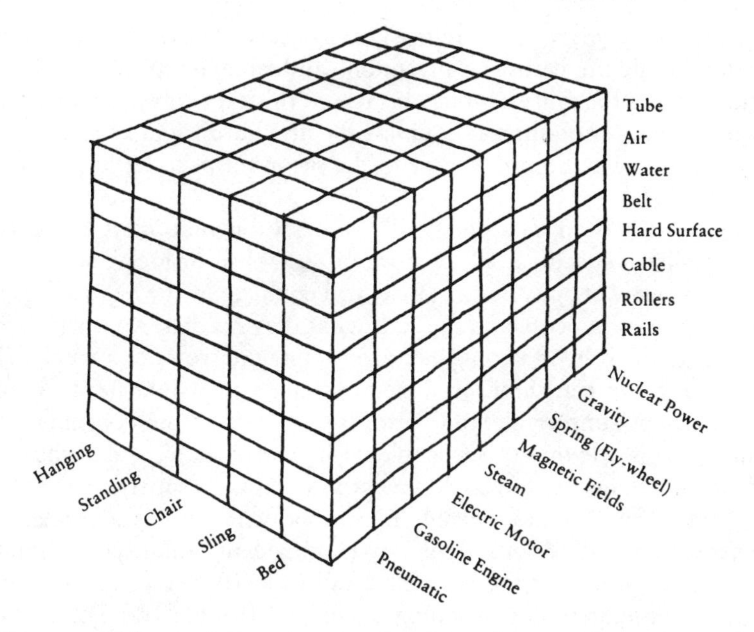

If we consider each box on the figure we have generated, we find that it represents a particular combination of our three parameters. For instance, one represents a steam-driven system that runs on rails and has passengers in chairs. This is not so interesting because it is a train and has already been thought of. So has the system that is driven by electricity and slings people from a cable (ski lift) and the gasoline engine powered one that seats people and travels on a hard surface. However, how about the pneumatic powered one in which people lie down and are transported through a tube, or the gravity powered one in which people stand and are transported down a belt? If one has access to a computer and can therefore consider large numbers of parameters, this technique can furnish enough combinations to keep the problem-solver well out of trouble as he sifts through them looking for something to spark an elegant solution to a problem. The creative "purist" would tend to scoff at this method as being too mechanistic. However, morphological analysis does, in fact, produce conceptual information.

Let me give you another example of the use of lists as thinking (rather than memory) aids. I am going to ask you to make a "bug list." People with a healthy fantasy life often play with the concept of inventing something the world needs and retiring on the proceeds. However, relatively few of them accomplish this. There are two factors that explain this lack of follow-through. The first is the difficulty in thinking of something specific that the world needs. The second is that it may require many years of apprehension, financial deprivation, and floundering family life before an invention can be made to pay off. The second factor is the more serious obstacle to this type of retirement. However, since it is not important unless the first hurdle can be cleared, and since we have no solutions here to the second aspect, let us pursue the first further.

In order to think of a potentially successful invention, it is necessary to establish a specific need. One way to establish or locate such a need is to interview people. For instance, you could go to the nearest hospital and start asking people on the staff what they needed. Another method is to play the role of a consumer group. Imagine you are a truck driver and see if you can think of something that you would need. A third, and perhaps simpler method, is to use yourself as the consumer. You must have needs that other people in the world share, and if you could identify such needs, you could invent something to satisfy them.

A problem that most people must cope with here is a tendency to generalize. If one of your needs is to eliminate air pollution or eliminate violence, you are setting yourself a tall task. (Better you have a need such as eliminating dog droppings from your front lawn.) The best way of starting on your retirement is probably to come up with a list of specific small-scale needs. A bug list is such a list. It contains as fluent and flexible and as specific and personal a list as possible of things that bug you.

> **Exercise:** Take a paper and pencil and construct such a list. Remember humor. If you run out of bugs before 10 minutes, you are either suffering from a perceptual or emotional block or have life unusually under control. If you cannot think of any bugs, I would like to meet you.

On the next page is a list of bugs from a few present-day Stanford student lists. After you are done see if your bugs are as flexible, specific, and personal as theirs. (You may also feel free to draw any conclusions you care to concerning students from this list on your own time.)

If properly done, your bug list should spark ideas in your mind for inventions. The list should ensure that specific areas of need are illuminated and that you have put in a reasonable amount of fluency and flexibility of thought. It should contain far-out bugs as well as common ones. For many of you, it may be the most specific thinking you have ever done about precisely what small details in life bother you.

After our students make such lists, we often ask them to turn them into inventions. Almost invariably, an interesting "invention" results. This requires first of all that the list be reduced to a few bugs of more than average potential. (Some preliminary thinking of solutions may occur here. Needs always seem to have more potential if a clever solution is available). Next we ask them to produce several concepts for the solution of each chosen bug. We then ask them to choose a concept and work it out in detail (including physical designs, where pertinent, and implementation plans). You may try this process if you would like. If you succeed and make a lot of money, send me some of it and I will use it for a charitable cause (my house payments).

Bug List

TV dinners
buying a car
relatives
paperless toilets
men's fashions
rotten oranges
hair curlers in bed
hypodermic needles for shots
sweet potatoes
cleaning the oven
no urinals in home bathrooms
bumper stickers that cannot be removed
broken shoe-laces
ID cards that don't do the job
pictures that don't hang straight
ice cubes that are cloudy
glary paper
swing-out garage doors
dripping faucets
doors that swell and stick in damp weather
newspaper ink that rubs off
bikes parked in wrong place
lousy books
blunt pencils
burnt out light bulbs
panty hose
thermodynamics
dirty aquariums
noisy clocks
plastic flowers
instant breakfast
buttons which must be sewn
hangnails
small, yapping dogs
waste of throw-away cans
soft ice cream
crooked cue sticks
prize shows on TV
static charges—car, blankets, etc.
ditches for pipe that are dug too large
bathtubs
cigarettes
balls which have to be pumped up
changing from reg. to sunglasses
reading road map while driving
wobbly tables and chairs
big bunches of keys
shoe heels that wear out
campers that you can't see around
corks that break off in wine bottles
soap dishes that you can't get the soap out of
vending machines that take your money with no return
buzzing of electric shavers
pushbutton water taps
Presto logs
one sock
stamps that don't stick
chairs that won't slide on the floor
banana slugs
trying to get change out of pockets
red tape
smelly exhausts
high tuition
writing letters
strip mining
dull knives
conversion of farm land to homes
chlorine in swimming pools
polishing shoes
broken spokes
stripped threads
cold tea
X-rated movies that shouldn't be X-rated
bras
mowing lawns
locating books in library
miniature poodles
parents' deciding a kid's career
solicitors—telephone and door-to-door
typewriter keys sticking
shock absorbers that don't work
shaving

Another type of list is the "check list." This is a list that you can apply to the thinking process to make sure that you have not been trapped by blocks. John Arnold, founder of the Design Division at Stanford and one of the pioneers in design education, used a check list first put forth by Alex Osborn in his book *Applied Imagination.* It is reproduced below.

Check List for New Ideas

Put to other uses?
New ways to use as is? Other uses if modified?

Adapt?
What else is like this? What other idea does this suggest? Does past offer a parallel? What could I copy? Whom could I emulate?

Modify?
New twist? Change meaning, color, motion, sound, odor, form, shape? Other changes?

Magnify?
What to add? More time? Greater frequency? Stronger? Higher? Longer? Thicker? Extra value? Plus ingredient? Duplicate? Multiply? Exaggerate?

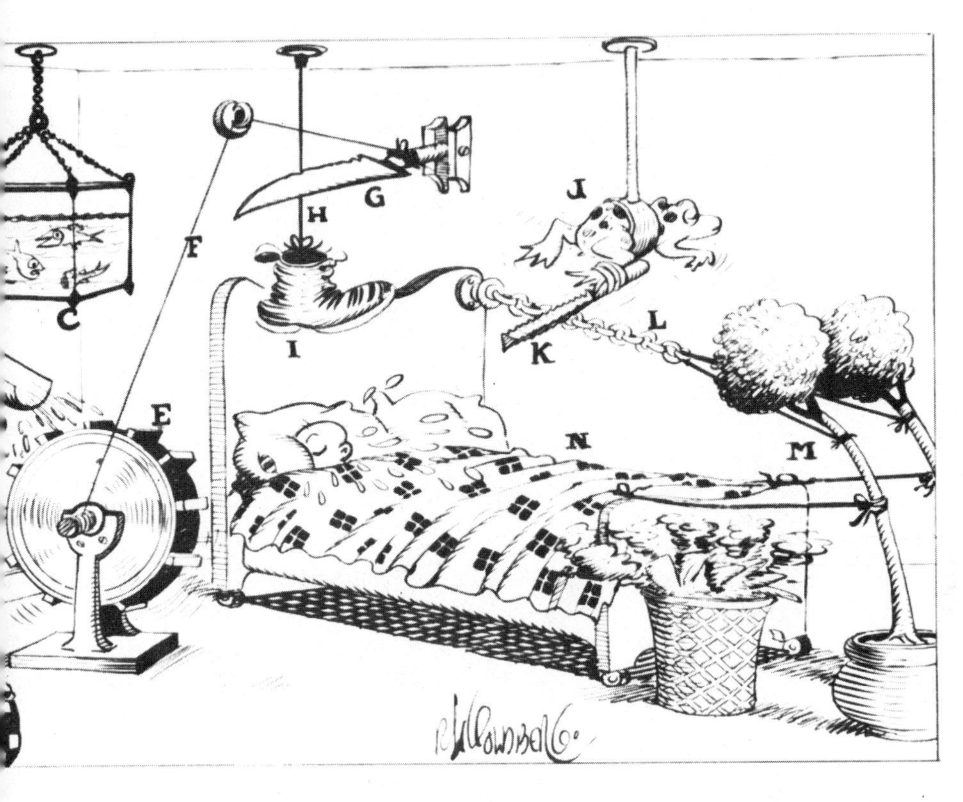

Minify?
What to subtract? Smaller? Condensed? Miniature? Lower? Shorter? Lighter? Omit? Streamline? Split up? Understate?

Substitute?
Who else instead? What else instead? Other ingredient? Other material? Other process? Other power? Other place? Other approach? Other tone of voice?

Rearrange?
Interchange components? Other pattern? Other layout? Other sequence? Transpose cause and effect? Change pace? Change schedule?

Reverse?
Transpose positive and negative? How about opposites? Turn it backward? Turn it upside down? Reverse roles? Change shoes? Turn tables? Turn other cheek?

Combine?
How about a blend, an alloy, an assortment, an ensemble? Combine units? Combine purposes? Combine appeals? Combine ideas?

Professor Arnold had this check list put on a deck of cards, which he would ruffle through to see whether any of them would extend his thinking.

Koberg and Bagnall suggest in *The Universal Traveler* a list of what they call "manipulative" verbs—Osborn's original check list could be augmented by adding such words as:

Multiply	Distort	Fluff-up	Extrude
Divide	Rotate	By-pass	Repel
Eliminate	Flatten	Add	Protect
Subdue	Squeeze	Subtract	Segregate
Invert	Complement	Lighten	Integrate
Separate	Submerge	Repeat	Symbolize
Transpose	Freeze	Thicken	Abstract
Unify	Soften	Stretch	Dissect, etc.

Another type of check list is the one below, developed by George Polya of Stanford for use in solving single-answer mathematical problems. It first appeared in his book *How to Solve It*. This list not only exercises questioning ability, but also your fluency, flexibility, and originality through increased observation and association.

Understanding the Problem

What is the unknown? What are the data? What is the condition? Is it possible to satisfy the condition? Is the condition sufficient to determine the unknown? Or is it insufficient? Or redundant? Or contradictory? Draw a figure. Introduce suitable notation. Separate the various parts of the condition. Can you write them down?

Devising a Plan

Have you seen it before? Or have you seen the same problem in a slightly different form? Do you know a related problem? Do you know a theorem that could be useful? Look at the unknown! Try to think of a familiar problem having the same or a similar unknown.

Here is a problem related to yours and solved before. Could you use it? Could you use its results? Could you use its method? Should you introduce some auxiliary element in order to make its use possible? Could you restate the problem? Could you restate it still differently? Go back to definitions.

If you cannot solve the proposed problem try to solve first

> some related problem. Could you imagine a more accessible related problem? A more general problem? A more special problem? An analogous problem? Could you solve a part of the problem? Keep only a part of the condition, drop the other part; how far is the unknown then determined, how can it vary?
>
> Could you derive something useful from the data? Could you think of other data appropriate to determine the unknown? Could you change the unknown or the data, or both, if necessary, so that the new unknown and the new data are nearer to each other? Did you use all the data? Did you use the whole condition? Have you taken into account all essential notions involved in the problem?
>
> **Carrying Out the Plan**
>
> Carrying out your plan of the solution, check each step. Can you see clearly that the step is correct? Can you prove that it is correct?
>
> **Examining the Solution Obtained**
>
> Can you check the result? Can you check the argument? Can you derive the result differently? Can you see it at a glance? Can you use the result of the method for some other problem?

List-making techniques can be used by anyone to assemble alternate concepts. They apply to the most rigid thinker as well as the most playful. They not only ensure good definition, but also that the ideas will last, since they are committed to paper. As we have already mentioned, ideas beget other ideas. If they are listed, they will lie around for days goading the idea-haver into other thoughts.

In a sense, such techniques as design notebooks, idea books, and problem journals are list-making techniques. A chronological record of a problem-solution is a list of all of the thoughts that have occurred during the solution of the problem. By its very existence, it causes the problem-solver to have more and increasingly imaginative concepts, especially if occasionally reviewed by others. We ask most of our students to keep design notebooks during their project work. These notebooks are complete chronological records of the thinking they have done and the information they have acquired. We collect them periodically for grading. We know that many of the students consider the notebooks to be an odious task and of questionable value and make most of their entries the night before we collect them. The impact this last minute "keeping up to date" has on their thinking is obvious. Groups that are apparently stuck on a problem will magically come up with new approaches the day that they hand in their notebooks. The intensive list-making they go through when padding up their notebook to meet our expectations is a powerful stimulus to conceptualization.

Other "conscious" blockbusters can be found in almost any "how to do it" book on creativity. Several of these are mentioned in the Reader's Guide at the end of the book. Most of them utilize some degree of list-making. Most of them include some gimmick to encourage playful thinking without the requirement of confronting playfulness head-on. Most of them work, although effort is required to utilize them in a realistic problem-solving situation. The tendency of most teachers and authors (myself included) when trying to show the power of a technique is to include a sample problem which is a set-up for demonstrating it. Usually it is a simple problem that can be instantly solved if the technique is used and only with difficulty if it is not (e.g., visualization—the monk puzzle). The game becomes more difficult when you leave sample problems.

However, if you acquire sufficient practice with conscious blockbusters, they can be applied to complex "real" problems quite successfully. In fact, after sufficient usage, they will become second-nature. The specific listing of conceptual blocks is a conscious blockbusting technique. If specific examples are furnished, it is seemingly easy to gain an appreciation for these blocks. However, it is harder to identify them in one's own thinking, both because they are blocks and because one's thinking is usually more complex than the examples. However, if you put a good deal of conscious effort into looking for them, you will learn to identify them. You will learn what types of blocks to expect in various situations, aggressively search them out, and gleefully violate them.

Unconscious Blockbusting

In Chapter Three, we discussed the critical role played by the unconscious mind and its inhibition by the ego and superego. So far in this chapter, we have discussed techniques whereby you can consciously force your way through conceptual blocks. Through the use of various forms of listing and by consciously questioning and striving for fluency and flexibility of thought, it is possible to improve considerably your conceptual performance. These techniques work by utilizing the intellectual problem-solving capability of the conscious mind. In the remainder of this chapter, we will be concerned with decreasing the inhibiting effect of the ego and superego on the unconscious mind.

One of the most powerful techniques of enhancing your conceptual ability is the postponement of judgment mentioned in Chapter Four. The ego and superego suppress ideas by judging them to be somehow out of order as they try to work their way up to the conscious level. If this judging can be put aside for a while, many more ideas will live until they can be "seen." The dangers of premature judgment are alluded to in the following statement by Schiller (taken from a personal letter to a friend and contained in *The Basic Writings of Sigmund Freud*, edited by A. A. Brill).

> The reason for your complaint [about not being creative] lies, it seems to me, in the constraint which your intellect imposes upon your imagination. Here I will make an observation, and illustrate it by an allegory. Apparently, it is not good—and indeed it hinders the creative work of the mind—if the intellect examines too closely the ideas already pouring in, as it were, at the gates. Regarded in isolation, an idea may be quite insignificant, and venturesome in the extreme, but it may acquire importance from an idea which follows it; perhaps, in a certain collocation with other ideas, which may seem equally absurd, it may be capable of furnishing a very serviceable link. The intellect cannot judge all those ideas unless it can retain them until it has considered them in connection with these other ideas. In the case of a creative mind, it seems to me, the intellect has withdrawn its watchers from the gates, and the ideas rush in pell-mell, and only then does it review and inspect the multitude. You worthy critics, or whatever you may call yourselves, are ashamed or afraid of the momentary and passing madness which is found in all real creators, the longer or shorter duration of which distinguishes the thinking artist from the dreamer. Hence your complaints of unfruitfulness, for you reject too soon and discriminate too severely.

Delaying judgment does not come easily to most people, since we are taught to be severe critics of anything impractical, unrealistic, flippant, flawed, or socially frowned upon. Often we do not want to admit, even to ourselves, the existence of such thoughts in our mind. We certainly do not want to admit to others that we might think of roofing a building with feathers, of reducing air pollution by substituting sedan chairs for automobiles, or perhaps even of legalizing heroin to reduce crime. However, our minds should be able to conjure up these and much wilder ideas if we are to be truly creative thinkers. How, then, do we delay judgment? We can begin by using the conscious mind. Often if we can consciously make our ego relax a little, the success of the idea generation that follows may cause it to relax even further. We begin a game that is to some extent self-perpetuating. The easiest way to begin this game is to formally (by agreement with oneself or with others) establish a judgment-suspension session. Individually I may say to myself, "All right, I need some fresh ideas on this problem I am working on and I have a little time to spend, so I will suspend judgment and see what ideas I can think of. It doesn't matter if my thoughts are weird at times, since no one can see what I am up to." I am then free to conceptualize without judging the practicality of the ideas, since I am not imperiling my ego. After all, I officially announced to myself that I would undergo this activity and therefore it is not typical of my usual mental deportment.

Suspending judgment in groups can be even more effective than suspending it individually, since a spirit of enthusiasm can develop in the group and ideas may spark ideas in others. An extremely well-known technique for achieving this is brainstorming, which is described in detail in Chapter Eight. Another interesting technique that encourages suspension of judgment was developed in the early days of Synectics Inc. in Cambridge, Massachusetts. It is described in detail in *Synectics* by William J. J. Gordon.

This technique utilizes metaphor. Four types of operational mechanisms are used: Personal Analogy, Direct Analogy, Symbolic Analogy, and Fantasy Analogy. The *personal analogy* requires that the problem-solver identify with part or all of the problem and its solution. The *direct analogy* attempts to solve a problem by the direct application of parallel facts, knowledge, technology, or whatever. The *symbolic analogy* is somewhat like the personal analogy, except that the identification is between the problem and objective and impersonal objects or images. The *fantasy analogy* allows the problem-solver to use fantasy to solve the problem.

The best way of showing the use of the operational mechanisms is to quote a passage from Gordon's book. This is a so-called Synectics excursion between five people faced with the problem of inventing a vapor-proof closure for space suits. As this passage begins the group has just finished discussing the question, "How do we in our wildest fantasy desire the closure to operate?"

Fantasy Analogy

G: Okay. That's over. Now what we need here is a crazy way to look at this mess. A real insane viewpoint . . . a whole new room with a viewpoint!

T: Let's imagine you could will the suit closed . . . and it would do just as you wanted by wishing . . . (Fantasy Analogy mechanism)

G: Wishing will make it so . . .

F: Ssh, Okay. Wish fulfillment. Childhood dream . . . you wish it closed, and invisible microbes, working for you, cross hands across the opening and *pull* it tight . . .

B. A zipper is kind of a mechanical bug (Direct Analogy mechanism). But not airtight . . . or strong enough . . .

G: How do we build a psychological model of "will-it-to-be-closed"?

R: What are you talking about?

B: He means if we could conceive of how "willing-it-to-be-closed" might happen in an actual model—then we . . .

R: There are two days left to produce a working model—and you guys are talking about childhood dreams! Let's make a list of all the ways there are of closing things.

F: I hate lists. It goes back to my childhood and buying groceries . . .

R: F, I can understand your oblique approach when we have time, but now, with this deadline . . . and you still talking about wish fulfillment.

G: All the crappy solutions in the world have been rationalized by deadlines.

T: Trained insects?

D: What?

B: You mean, train insects to close and open on orders? 1-2-3 Open! Hup! 1-2-3 Close!

F: Have two lines of insects, one on each side of the closure—on the order to close they all clasp hands . . . or fingers . . . or claws . . . whatever they have . . . and then closure closes tight . . .

G: I feel like a kind of Coast Guard Insect (Personal Analogy mechanism).

D: Don't mind me. Keep talking . . .

G: You know the story . . . worst storm of the winter—vessel on the rocks . . . can't use lifeboats . . . some impatient hero grabs the line in his teeth and swims out . . .

B: I get you. You've got an insect running up and down the closure, manipulating the little latches . . .

G: And I'm looking for a demon to do the closing for me. When I will it to be closed (Fantasy Analogy mechanism), Presto! It's closed!

B: Find the insect—he'd do the closing for you!

R: If you used a spider . . . he could spin a thread . . . and sew it up (Direct Analogy mechanism).

T: Spider makes thread . . . gives it to a flea . . . Little holes in the closure . . . flea runs in and out of the holes closing as he goes . . .

G: Okay. But those insects reflect a low order of power . . . When the Army tests this thing, they'll grab each lip in a vise one inch wide and they'll pull 150 pounds on it . . . Those idiot insects of yours will have to pull steel wires behind them in order . . . They'd have to stitch with steel. *Steel* (Symbolic Analogy mechanism).

B: I can see one way of doing that. Take the example of that insect pulling a thread up through the holes . . . You could do it mechanically . . . Same insect . . . put holes in like so . . . and twist a spring like this . . . through the holes all the way up to the damn closure . . . twist, twist, twist, . . . Oh, crap! It would take hours! And twist your damn arm off!

G: Don't give up yet. Maybe there's another way of stitching with steel . . .

B: Listen . . . I have a picture of another type of stitching . . . That spring of yours . . . take two of the . . . let's say you had a long demon that forced its way up . . . like this . . .

R: I see what he's driving at . . .

B: If that skinny demon were a wire, I could poke it up to where, if it got a start, it could pull the whole thing together . . . the springs would be pulled together closing the mouth . . . Just push it up . . . push—and it will pull the rubber lips together . . . Imbed the springs in rubber . . . and then you've got it stitched with steel!

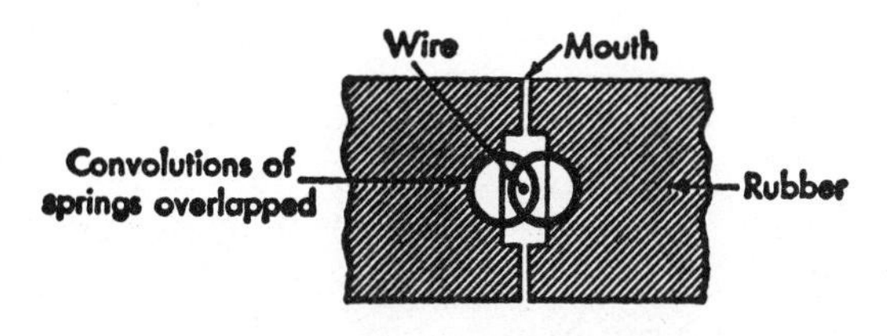

Cross-sectional Diagram

The Synectics technique results in a group of people delaying judgment in that they are willing to entertain ideas which normally they would probably reject as impractical. They can do so because the Synectics rules tell them to. They are effectively using a technique that, temporarily at least, relaxes the watchdog function of the ego and superego and lets the subconscious mind come forth with concepts. These techniques can be used individually as well as within a group, although the synthesis effect of the group will not enter into play. If you happen to think of using a brick as a hot water bottle, the realization that you are making use of the thermal capacity of the brick should quickly spark similar uses (bed warmer, warm floors, luau pit) in your mind just as it would in a group.

But what about ways to alter the ego's regulation of the unconscious mind without such techniques? Is it possible to somehow relax the control of the ego and superego in general, so that one may make better use of the unconscious mind? The answer is yes, but it is not simple. Let me try to explain the reasons for the difficulty by returning briefly to a discussion of psychology.

As mentioned previously, although there are many theories on the conceptual process to explain various characteristics of conceptualization, complete understanding does not exist. Many psychologists have attempted to explain the mental processes needed in creative thought and the motivations and characteristics of creative people.

Maslow

Abraham Maslow was one of the more significant figures in the attempt to understand creativity. He preferred to talk about "primary" and "secondary" processes rather than about the unconscious and the conscious. Rather than referring to the primary mental process as the unconscious, he called it the "deeper" self, in the sense that it is hidden beneath the surface. Maslow also discussed primary and secondary creativity, as mentioned in Chapter Three. Secondary creativity is that which is evidenced by most people working in a system which requires a great amount of "discipline" work. It utilizes "right-handed" thinking and is based upon breakthroughs (primary creativity) made by others.

To Maslow, primary creativity comes from the deeper or primary self. It is common and universal in children, but in many adults is blocked off to a great extent. In an address called "Creativity in Self-Actualizing People" (contained in *Creativity and Its Cultivation*, edited by Harold Anderson), Maslow discussed a study he did of people who were especially creative and, as he puts it, "self-actualizing." He prefaces his discussion with the admission that very early he abandoned the notion that health, genius, talent, and productivity went together. Maslow found that a great number of people he studied were highly creative in terms of their own self-actualizing capacity, yet were neither particularly productive nor possessors of great "talent" or "genius." He was particularly struck by those people among his subjects who, though they had no noteworthy talent in any area conventionally associated with creativity, were in their daily lives original, novel, ingenious, and inventive. From this, Maslow says he learned to apply the word "creativeness" to many activities, processes, and attitudes other than the standard categories to which the quality is typically ascribed (such as literature, art, theories, etc.). Thus evolved Maslow's distinguishing between "special talent creativeness" (which is typically associated with creativity), and what he calls "self-actualizing creativeness" (primary creativity), which can be manifest in anything we do—in our most ordinary, mundane activities. In studying the traits of self-actualizing people who bring to their everyday affairs an attitude and manner of creativeness, Maslow located certain commonalities. He found these people to be more spontaneous, expressive, and natural and less controlled and inhibited in their behavior than the average. Their behavior seemed less blocked and less self-critical: "This ability to express ideas and impulses without strangulation and without fear of ridicule from others turned out to be an essential aspect of self-actualizing creativeness." Maslow found his subjects to be different from the average person in another way which he felt made creativity more likely: "Self-actualizing people are relatively unfrightened by the unknown, the mysterious, the puzzling, and often are positively attracted by it; i.e., selectively pick it out to puzzle over, to meditate on, and to be absorbed with."

Maslow saw a connection between *creativity in one's actions* and the *inner integration of one's self:* "To the extent that creativeness is constructive, synthesizing, unifying, and integrative, to that extent does it depend in part on the inner integration of the person." Maslow traces this to "the relative absence of fear" found in these subjects—both of others (what they would say, laugh at, demand) and especially of themselves (their insides, impulses, emotions, thoughts). "It was this approval and acceptance of their deeper selves that made it possible to perceive bravely the real nature of the world and also made their behavior more spontaneous (less controlled, less inhibited, less planned, less 'willed,' and designed). By contrast, average and neurotic people walled off, through fear, much that lay within themselves. They controlled, they inhibited, they repressed, and they suppressed. They disapproved of their deeper selves and expected that others did, too." By doing this, Maslow explains, the person "loses a great deal, too, for these depths are also the source of all his joys, his ability to play, to love, to laugh, and, most important for us, to be creative."

Barron

Frank Barron, another psychologist who has done a great deal of research on creativity, takes a slightly different approach. He is unwilling to accept overall psychological health as the criterion for a creative person, because he feels it necessary to formulate criteria which admit to such creative talents as Beethoven, Hooke, Swift, Van Gogh, Rimbaud, Baudelaire, Bronte, Heine, Wagner, and others who created unhappiness. In an article in *Scientific American*, September 1958, Barron states:

> I would propose the following statements as descriptive of creative artists, and perhaps also of creative scientists:
>
> Creative people are especially observant, and they value accurate observation (telling themselves the truth) more than other people do.
>
> They often express part-truths, but this they do vividly; the part they express is the generally unrecognized; by displacement of accent and apparent disproportion in statement they seek to point to the usually unobserved.
>
> They see things as others do, but also as others do not.
>
> They are thus independent in their cognition, and they also value clearer cognition. They will suffer great personal pain to testify correctly.
>
> They are motivated to this value and to the exercise of this talent (independent, sharp observation) both for reasons of self-preservation and in the interest of human culture and its future.
>
> They are born with greater brain capacity; they have more ability to hold many ideas at once, and to compare more ideas with one another—hence to make a richer synthesis.
>
> In addition to unusual endowment in terms of cognitive ability, they are by constitution more vigorous and have available to them an exceptional fund of psychic and physical energy.
>
> Their universe is thus more complex, and in addition they usually lead more complex lives, seeking tension in the interest of the pleasure they obtain upon its discharge.
>
> They have more contact than most people do with the life of the unconscious, with fantasy, reverie, the world of imagination.
>
> They have exceptionally broad and flexible awareness of themselves. The self is strongest when it can regress (admits primitive fantasies, naive ideas, tabooed impulses into consciousness and behavior), and yet return to a high degree of rationality and self-criticism. The creative person is both more primitive and more cultured, more destructive and more constructive, crazier and saner, than the average person.

Others have hypothesized other models of the creative person, ranging from the happy, well-balanced, suntanned, confident extrovert to the pain-riddled, warped, moody neurotic. Yet, the theme of unguarded unconscious (preconscious or primary self or whatever term we use) surfaces again and again. Therefore let us ask the question: What can we do to free the unconscious from its over-zealous warden?

Psychoanalysis might be an obvious thought, for it is intended to better integrate a personality by making the unconscious more conscious. It supposedly can ameliorate compulsive-obsessive behavior and therefore unlock the primary self. Many psychologists who have studied creativity agree that the goals of psychoanalysts, if reached, should enhance creativity. The fear that psychoanalysis will somehow "ruin" the creative powers of a person has been dismissed by most of the experts. However, for most people, psychoanalysis appears to be somewhat strong medicine for the improvement of creativity. It is expensive, it takes a long time, and its success is not predictable. Psychoanalysis may be attractive if one's behavior is such that life has become acutely unpleasant or unbearable. However, for "normal neurotics" it is perhaps overkill if we're interested in the enhancement of creativity.

Other Paths for Freeing the Unconscious

Are there other ways to free the unconscious? Probably many. Maslow feels that any technique that increases self-knowledge should in principle increase creativity. In the cultures of the Middle and Far East there have existed for many hundreds of years what Robert Ornstein, in his book *The Psychology of Consciousness*, calls "The Traditional Esoteric Psychologies." These have been concerned with personal, empirical approaches to self-knowledge, rather than with impersonal scientific approaches. These psychologies have often developed within disciplines such as Buddhism or Yoga and have utilized techniques such as meditation which are specifically designed to temporarily minimize linear logical thought and strengthen certain mental processes ascribed to the unconscious. Science is just beginning to understand "mystical experiences" and different "levels of consciousness." Ornstein believes that such experiences may be instances where the analytical left side of the brain relinquishes its usual control of consciousness and enables the right side to more freely interpret stimuli in a non-linear, non-deductive way. In his book, Ornstein makes a compelling argument for the integration of such psychologies and techniques with the "right-handed" psychology with which we are familiar. Certainly a higher degree of self-knowledge would result along with an increased respect for "left-handed" thinking. However, these techniques, as practiced in the Middle and Far East, take time and effort, and though they are currently becoming increasingly popular in the Western world, many of us are far from being able and/or willing to use them to improve our conceptual ability.

What paths are easily available to allow us right-handed Westerners to better free the unconscious? Maslow suggests education as one, and I, as an educator rather than a psychiatrist or mystic, must heartily agree. Maslow suspects that although education does little for relieving the repression of "instinct" and "forbidden impulses," it is quite effective in integrating the primary processes and conscious life. Knowledge about the psychological processes, about problem-solving, and especially about one's self can loosen the control of one's ego. The principle involved is a simple one: things are not as threatening when they are understood. Fears are lessened if their sources are understood, and most people's egos are "smart" enough to relax a bit if they are convinced that the results may be positive.

Reading is one of the best ways to gain such self-knowledge. Many books and articles are available concerning the strengthening of one's self-esteem and self-confidence and freeing oneself from unnecessary fears and insecurities. The Reader's Guide section at the back outlines a number of starting points should you desire to journey further, with references to creativity research, psychological theory, and "self-therapy." The books on creativity research will give you a better overview of what is known about creative thinking and the characteristics of highly creative people. Psychological theory books will help you understand human behavior and how the mind works. The "self-therapy" books seek to apply psychological theory in a way in which you can affect your own behavior.

Understanding the workings of your mind is somewhat like understanding a golf swing. It allows you to work on changing your present actions in a detailed and conscious way. However, in the case of creative thinking, a side benefit is achieved in gaining a greater understanding of the workings of other people's minds as well. Many fears demand a comparison with other people for their maintenance. The fear of asking questions is often predicated on exposing your ignorance to others. The fear goes away when you realize that others are ignorant too. Similarly, you are less afraid of expressing your emotions when you learn that others have similar emotions, whether they have repressed them or not. Brainstorming works because the other people have silly ideas too. You are more willing to struggle with a problem when it is realized that few people consistently give birth to answers or solutions in a blinding flash of pure inspiration.

Therefore I encourage you to read. The sport of thinking about thinking is an interesting one, and the literature to help you in this pastime is extensive. It can only lead to a better ability to use your own mind—a thorough knowledge of psychological theory and creativity research cannot help but increase your creativity. I feel that we have some effect on the creativity of our students merely by making a "big deal" out of creativity. By elevating it to the status of a class subject and by thus bringing it out of the underground, we cause our students to attach the same importance to it as they do to their other academic subjects and therefore to feel that they should, in fact, be more creative.

Reading, of course, is not the only way to gather knowledge about creativity and conceptualization. You may talk to psychologists and psychiatrists who are involved in trying to give people freer access to their unconscious mind. You may observe those about you and attempt to correlate their actions and their thought processes to their creative output. You may become more introspective about your own thinking (a must in any case) in an attempt to make it more creatively powerful and efficient.

One of the most important activities you should engage in is trying to free your unconscious to engage in creative thinking. If you brainstorm (or synect) or merely *consciously force yourself to be creative* (by use of lists or whatever), a strange thing happens. First of all, you usually find that if anything you are more successful in the world, rather than less (so some of your fears were groundless). You also find that creative thinking comes easier to you. There are psychological reasons for this. If you use your unconscious level, your consciousness gets the message that such activities are all right. This message is strongly reinforced if some of the outputs from the unconscious result in successes which the ego can revel in. The more creative thinking is done, the more natural and rewarding it becomes and the more the ego relaxes.

Photograph G.D. Hackett, New York.

CHAPTER EIGHT

GROUPS AND ORGANIZATIONS

So far, we have considered only indirectly the effects of other people on individual thinking. However, much problem-solving takes place in group and oranizational settings. We conceptualize with family members, friends, coworkers in community groups and volunteer committees, and with professional colleagues. In such situations we directly affect other people's conceptual process and they directly affect ours.

It should not surprise anyone that the conceptual process can suffer when many people are involved. "Group think" is hardly a phrase of acclaim; it implies blandness and lack of creativity. Most of us have heard the venerable definition of a camel as a "horse designed by a committee." But we must also realize that groups and organizations can excel in the conceptual process. Groups of people can bring many diverse perceptions and intellectual specialities to bear on a problem. They can provide a supportive emotional environment and the resources necessary to develop initial concepts into believable detail in a reasonable time.

In this chapter we will discuss the conceptual process in groups and organizations. We will first examine some of the possible blocks to conceptualization in groups of small enough size (from two to twenty members) that a great deal of formal structure is not required. We will then discuss internal and external motivation, rewards, and knowledge of the problem-solving process as they affect the creativity of people working in groups and organizations. Finally, we will make some comments on conceptualization in organizations so large that formal structure is necessary.

Small Groups: Affiliation Needs

In order to fulfill its function, a group or organization must often operate like an individual. It must be able to find problems, think up possible solutions, and make decisions. It must also operate with a reasonable level of creativity: too much and it loses its stability, too little and it fails. However, a group or organization differs from an individual in that each concept, or action, causes a response within each individual member. It may elate some members, depress others, fill others with fear, and seem misguided to still others. A group or organization is in effect a minisociety which, in its need to operate as if it had a single mind, places great pressures upon its members.

Each member of a group or organization, being human, has strong affiliation and ego needs. Affiliation needs urge the individual to act so as to gain the social acclaim of the group: to be liked, respected, and valued. Many psychological experiments have shown the strength of these needs. In one classic experiment a researcher asked people in various groups to estimate which of three lines of different lengths was equal in length to a fourth line. Only one person in each group was a real experimental subject. The others were "shills" who had been instructed to reach erroneous conclusions. About one-third of the experimental subjects who went through this experience changed their initial correct judgment to agree with that of the "shills," even though the difference in line lengths was clearly discernible.

Another example of the strength of affiliation needs is provided in a passage in Professor Harold Leavitt's fine book *Managerial Psychology*. This passage asks you to assume that you are a member of a professional committee who arrives at a meeting with a strong position on the first item on the agenda. After some discussion, you become aware that the other members of the committee share an opinion that is very different from yours. Initially, the other members of the committee show interest in your position and honestly attempt to understand it. They also attempt to explain the validity of their stand to you, but you are sure of your position. As time goes by, the mood of the meeting begins to change. The other members grow impatient as they are unable to sway you to the majority viewpoint. You become aware that you are starting to be attacked, and your mouth begins to dry and your stomach tightens. They begin to accuse you of being hostile, of sticking to a position even though you cannot come up with "new" reasons, and of delaying discussion of more important matters by your reluctance to join the consensus. However, you feel that you must ethically stick to your point.

After an hour and a half of discussion, you are the focus of the group. All the other members are heatedly arguing with you and using everything they can think of to sway you, since this is a committee that likes to operate by consensus and you are keeping them from reaching the type of agreement they pride themselves on. Finally, one of the committee members turns to the chairman and proposes that the committee agree on the majority opinion and move on to other matters. At this point, you realize that you are to be suddenly cut out of the group, and in fact you are. The members turn their chairs and face back toward the chairman. As the chairman summarizes the reasoning for adopting the majority opinion, you occasionally protest points you consider absurd. However, except for occasional glares, you get no acknowledgment from the committee. You have been disaffiliated.

It is easy for most people to identify with the characer in Leavitt's passage. It does not feel good. Most of us have had experiences with being psychologically rejected by a group of people we care about because we do not accept the common judgment. It is no wonder that people will accept wrong line-lengths and majority opinions which they consider wrong.

Affiliation needs underlie many of the conceptual blocks discussed in earlier chapters. People will like you if you think the way they do. But to the extent you succeed in aligning your thoughts with those of others, you can add to your perceptual and intellectual blocks. Problem-solving groups often become tightly knit and often consist of people who respect each other a great deal. Affiliation needs are particularly strong in such a situation, and severe emotional blocks can result. No one wants to fail in front of respected peers. A problem-solving group plays a strong role in creating its own subculture and environment. Blocks appear if these are not supportive to conceptualization. When a new concept deviates from the group's consensus, the originator may feel tempted to modify or swallow it. "Group think" can result.

However, when channeled positively, affiliation needs can result in high motivation (the desire for outstanding group performance) and a high degree of support. A group that understands the conceptual process can motivate individual members to think creatively, support them in doing so, and provide the atmosphere of trust that is vital if members are to conceptualize freely.

Ego Needs

Ego needs at times may work at cross-purposes with affiliation needs. They urge an individual to influence others, to lead, to be significant, to be outstanding. Unfortunately, one of the easiest ways to be significant within a group is to be critical, but as we saw earlier, a critical approach can be highly detrimental to conceptualization.

Equally detrimental, however, are misdirected attempts, especially those of a leader (formal or informal), to influence others. Influence techniques that are the most satisfying to the ego are not always the most supportive to conceptualization. The use of authority, for example, is a classic Western method of influencing others, and it can be very gratifying to the person issuing orders. We have all experienced authority because it is widely used by parents. It has certain advantages: it is quick, and it requires a minimum of knowledge about those in subordinate positions.

The authoritative style of influence usually results in people's being told quite precisely what to do. Unfortunately, this can decrease the motivation to do anything else—a clear inhibition to creativity. Authoritative leaders tend to give their subordinates answers, not problems; in addition, they often inspire rebellion. A climate of mutiny is hardly ideal for maximizing the conceptual output of a person or group. Rather than overthrow the ruler, people can rebel by having no ideas at all. They can mentally "stop out," and cease to contribute productively and creatively.

By contrast, a collaborative style of influence can be relatively cumbersome, but it encourages conceptualization. Communication tends to be more informal; each member is encouraged and expected to contribute, and each feels a responsibility for the success of the group venture and therefore gains satisfaction of affiliation and ego needs from the problem-solving process.

Which approach do you think would elicit the best conceptual output from you, authority or collaboration? Since you are a thinking person (otherwise you would not be reading this book), you would probably choose collaboration, and you would be right. Unfortunately, since collaborative management techniques do not always satisfy the ego needs of managers as thoroughly as do authoritative techniques, they are often overlooked in our egocentric culture.

Brainstorming

The proper handling of affiliation and ego needs in problem-solving groups is so critical, in fact, that specific techniques have been developed

to deal with them. Let us examine briefly two of these techniques. Perhaps the best known is brainstorming, a group problem-solving method given its name by Alex Osborn, the founder of the advertising firm of Batten, Barten, Durstine, and Osborn. In brainstorming the group's goal is defined to be conceptualization. Affiliation needs motivate members to contribute to this goal, since to be accepted, one should conceptualize according to the rules. Ego needs are controlled by absolutely prohibiting criticism and allowing no dominating role for a leader. Brainstorming groups generally consist of from five to ten people who work on a specific problem. According to Osborn, four main rules govern their behavior.

The first rule is that no evaluation of any kind is permitted. Osborn's explanation is that a judgmental attitude will cause the people in the group to be more concerned with defending ideas than with generating them. His second rule is that all participants be encouraged to think of the wildest ideas possible. His thinking here is that it is easier to tame down than to think up, and by encouraging wild ideas, internal judgment in the minds of the individual participants can be decreased. Third, Osborn encourages quantity of ideas, both because quantity also helps to control our internal evaluation and because he feels that quantity leads to quality. The final rule is that participants build upon or modify the ideas of others because, in his words, "combinations or modifications of previously suggested ideas often lead to new ideas that are superior to those that sparked them."

The brainstorming process benefits from having one member of the group act as a recorder, since a listing of the ideas as they are developed ensures that the group has continual access to its output and that ideas are not lost. The recording method should ideally be large enough in scale so that ideas are easily readable by everyone in the group. Brainstorming is most effective when the problem to be solved is simple and can be well defined. Brainstorming is useful at all levels of problem-solving, from the original attempt to formulate broad concepts to the final detailed definition.

There are a variety of behavioral reasons for brainstorming's success as a problem-solving technique. A study group at Harvard that investigated brainstorming in the 1950's listed these:

1. Less inhibition and defeatism: rapid fire of ideas presented by the group quickly explodes the myth that the individual often casts up that the problem overwhelms him, and that he can't think of a new and different solution.
2. Contagion of enthusiasm.
3. Development of competitive spirit; everyone wants to top the other's idea.

Still, delay of judgment is probably the most important factor that makes brainstorming work.

Brainstorming has at times received a bad name because it has been credited with generating ideas that are both shallow and in questionable taste. It has also been heavily spoofed and is sometimes identified with weirdness rather than thoughtfulness. However, the brainstorming process has some solid advantages and, if used when appropriate, can be extremely effective. A brainstorming group allows the pooling of a great diversity of background. Shallowness of output is often due to inadequate information available to the group and poor subsequent judgment, not to the technique. Brainstorming initially progresses rapidly when it attacks a problem because it is able to utilize common solutions. However, after these are used up, the process becomes more difficult because the members must come up with new concepts. It is in this later period that the technique has the most value. If the session is allowed to stop when the original rush of enthusiasm dies down (due to increased difficulty in thinking of ideas) it will not live up to its potential. The most effective way to learn more about brainstorming is to experience it.

Exercise: Find a group of people and set up a brainstorming session on a problem that is easily stated in precise terms. Try to think of a problem that is important to all of you. If you cannot, try one of the following:

1. Invent some sort of function that would allow you to become friends with a few fascinating people whom you know of, but do not know personally.
2. Invent (in reasonable detail) a better way to divide a large (2,000 sq. ft. or so) room into smaller spaces which can be used by various groups. This is an ongoing problem in schools. The dividing system should be flexible (space sizes easily changed), cheap, and aesthetically pleasing.
3. Invent an astounding entrée for a far-out dinner party.
4. Invent a better way for handling road maps in a car.
5. Invent a Christmas greeting card you can mail to your friends that will impress them for all time and let you avoid mailing future cards.

Synectics

Another group problem-solving technique is one developed by synectics Inc. in Boston, Massachusetts.* It is more complex than brainstorming and more sophisticated in that it allows criticism and a higher level of technical expertise but in a manner that does not allow the ego needs of the participants to jeopardize the conceptual process. Like brainstorming, it establishes the group's goal as problem-solving and thereby gives

*The original Synectics group split into Synectics Educational Systems, headed by William J. J. Gordon, and Synectics Inc., under the direction of George M. Prince. The technique above was developed by Synectics Inc. (The technique described in Chapter Seven predates the split.)

the participants the opportunity to satisfy their affiliation needs by solving the problem.

In this Synectics process the group works with a client who has a problem, giving the client ample opportunity to provide input to the group. The client originally states the problem and selects ideas from those presented. One particularly interesting feature of the Synectics process is that the leader does not contribute directly to the problem solution. The leader is a facilitator and a recorder and cannot contribute ideas of his or her own—the idea being that he/she is thus prevented from satisfying ego needs at the expense of the process.

The Synectics approach is particularly concerned with aiding corporations in the innovation process. In doing so, it relies upon a wide variety of principles and techniques. If a new concept is needed, Synectics uses an "excursion" such as the one shown here:

1. Leader asks client to select a directional Goal/Wish for which he/she'd like to develop Innovative Possible Solutions.
2. Leader picks a key word (action, concept) from the Goal/Wish.
3. Leader asks group (including client) to think of an example of that key word (action, concept) from a world that is distant from the world of the problem. The leader chooses the world. The leader then writes up a list of the group's examples.
4. Leader asks group to forget about the problem and the Goal/Wish and focus on any of the listed examples, thinking about its associations, images it conjures up, etc. Group members are asked to note these down on individual pads.
5. Leader asks group to use all or part of their example material to develop an Absurd Idea (probably impractical, impossible, or illegal) that addresses the original Goal/Wish.
6. Leader asks group to develop second generation ideas from any one of the Absurd Ideas (extracting key principles and applying them in a more realistic fashion without diluting the innovation).
7. Leader asks client to pick an idea that has appeal from the second generation ideas.
8. Process proceeds with paraphrase back to the original problem and itemized response from the client.

However, a large part of the energy in a Synectics session is devoted to the dynamics of the problem-solving group. The illustration below is taken from an article by George Prince, the founder of Synectics Inc., and shows the types of actions within a group that he feels encourage and discourage creativity.

The tone of a Synectics session is quite different from that of a

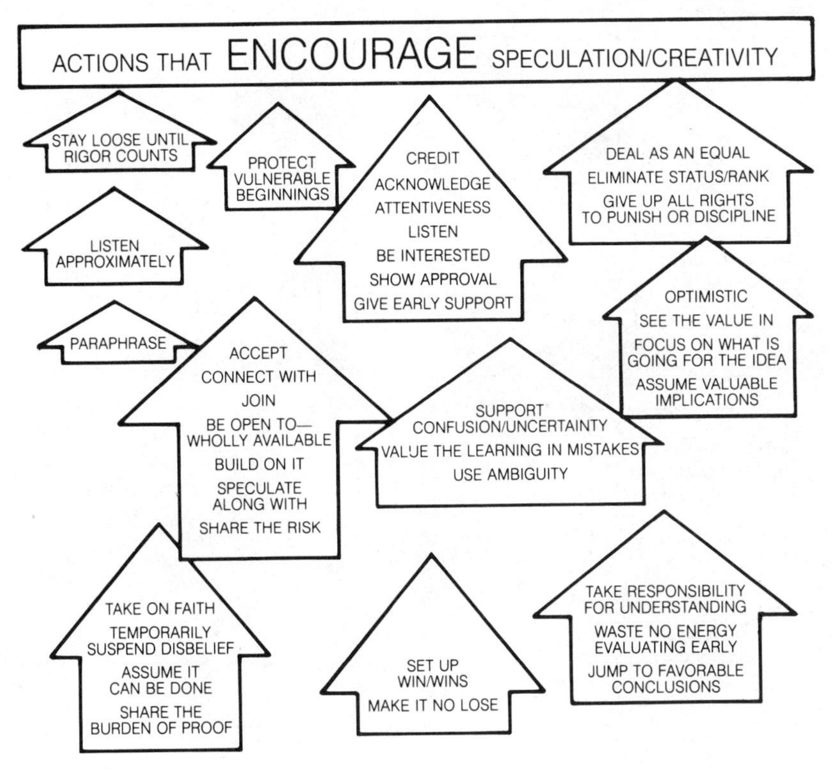

brainstorming session. In the brainstorming process, ideas fly and the participants satisfy their belonging needs and their ego needs by seeing how many imaginative ideas they can produce. In Synectics, fewer ideas are produced and belonging and ego needs are satisfied through helping the client solve his/her problem. In fact, one of the traits a Synectics leader must acquire is the ability to deal gracefully with group members whose ideas are not selected.

An interesting technique incorporated into Synectics sessions is the use of an approach to evaluation in which negative statements must be preceded by at least two positive statements. The positives serve to give the group continual indication of the desires of the client. However, they also reinforce the originator of the idea and help maintain a psychological atmosphere which is conducive to creativity. The criticism (the negative) is couched as a reservation rather than an overall no-vote and can immediately be made the problem for the next round. In this manner, rather strong criticism can be accommodated without inhibiting conceptualization.

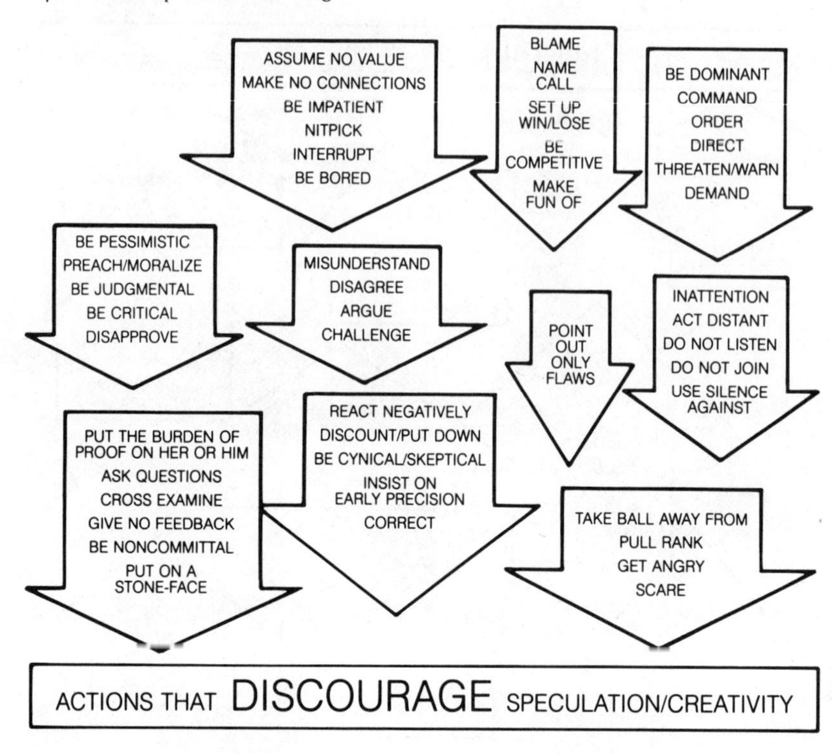

Techniques such as brainstorming and Synectics are effective in group problem-solving because they deal with affiliation and ego needs in ways that decrease conceptual blocks. However, a group can accomplish the same ends without such formal techniques, providing it understands the factors that inhibit the conceptual process. Such knowledge is especially important in the leader(s), whether leadership be formal or informal, since a group leader is in a position either to support or squelch conceptualization. However, conceptualization will flow even more freely if all members share this knowledge.

Intrinsic and Extrinsic Motivation

Let us now talk about the very important topic of intrinsic/extrinsic motivation and reward as a factor in encouraging creativity in groups. When thinking about creativity we cannot help but think of the exceptional figures of the past: Leonardo, Mozart, Einstein, et al. If we speculate on the reasons for the output of such people we find four recurring themes. First of all, they were unusual in their ability. They represented the extreme tail of the normal distributions which social scientists like to draw. Secondly, they were unusually resistant to the forces which society placed upon them. Although they obviously liked acclaim, they worked principally for their own satisfaction. Thirdly, they were either patronized by the wealthy, or otherwise found the wherewithal to support themselves while they devoted their primary effort to their pursuit. Finally, they all were involved in work which they loved. They were intrinsically motivated, in that they were motivated by their own interest in their work rather than by external factors.

If these generalizations are true, they are a bit worrisome to those of us interested in increasing creativity in groups and organizations. Do groups and organizations easily include those outside of the norm, especially if they are unusually immune to external inputs? Do they properly "patronize" them so that they can do exactly what they love?

There is convincing evidence that intrinsic motivation and the highest levels of creativity are closely related. An excellent (though technical) book entitled *The Social Psychology of Creativity*, by Teresa M. Amabile, makes this point in a most convincing way. (The book is also outstanding for its overall summarization of psychological theory pertinent to creativity.) The author has conducted experiments with children and adults on creative tasks such as collage, storytelling, poetry, and cartoon captioning and evaluated the results with a judging system which she demonstrates as reliable. She found that her subjects were most creative when motivated intrinsically. Extrinsic motivation factors (evaluation, peer observation, and rewards based on the quality of the output) decreased creativity. In her book, she supports her experimental conclusions with a myriad of references.

In one typical experiment, collages were constructed by 95 women enrolled in an introductory psychology course at Stanford University. They were not artists and had no significant previous experience in collage work. The collage was to convey the feeling of "silliness." The results were evaluated by fifteen artists who were shown to agree quite closely on their ratings.

The subjects were randomly divided into eight groups. Those in three of the groups were told that the only thing of interest was their mood and that the design itself was unimportant (no expectation of evaluation nor external motivation). The subjects in the first group were given no further focus, those in the second were asked to concentrate upon "technical goodness," and those in the third to concentrate upon creativity.

The subjects in the other five groups were told that the design would be evaluated by a panel of artists and that the quality of their collage would be part of the experimental data (evaluation expectation and external motivation). Once again those in the first group were given no further focus and those in the second were told that the judges would base their evaluation on "technical goodness." Those in the third group were told that they would be evaluated on how good their collages were technically, but were in addition given six detailed technical elements that the judges would consider. The subjects in the fourth group were told that the judges would base their evaluation on how creative the designs were, and those in the fifth group were not only told they would be evaluated on creativity, but were given seven specific criteria that the judges would consider.

The chart below shows the mean creativity for the collages assembled by the various groups. There was a major difference in the two groups that were given no focus. The subjects that were not expecting evaluation (intrinsic motivation) averaged much higher in creativity. The same is

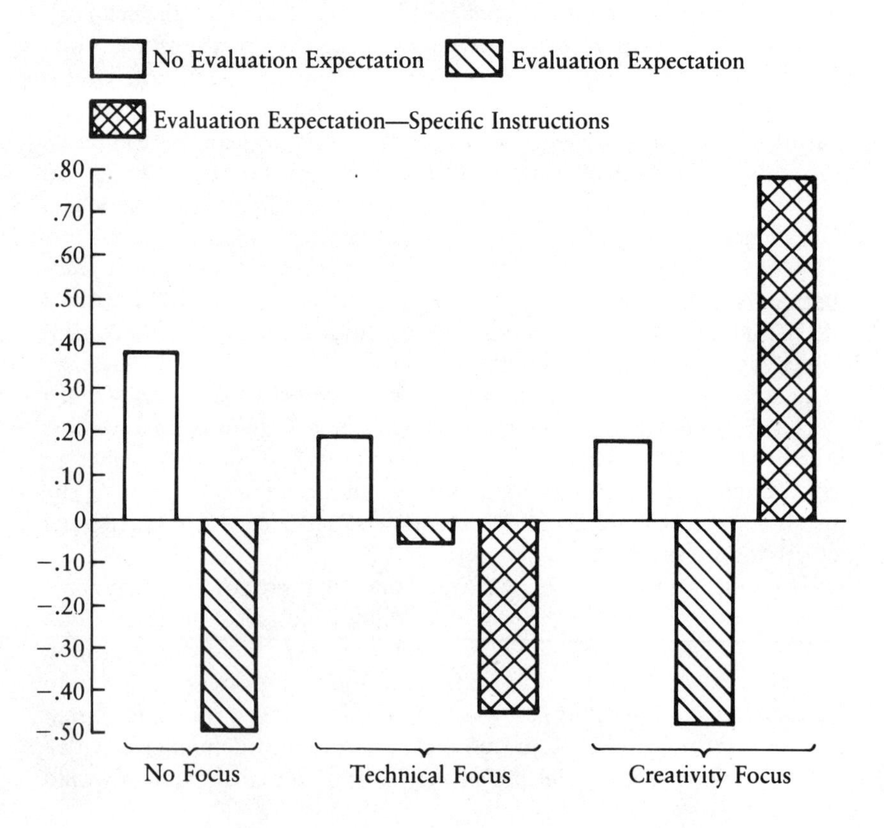

true of the three with the technical focus. Clearly, evaluation expectation also degraded the creativity of the subjects who were simply asked to focus on creativity. However, interestingly enough, those told the seven specific creativity criteria were judged to be the most creative of all. A bit more about that later, since it is a finding which is consistent with the message in this book.

From such data it would appear that we are more creative when intrinsically motivated, when working in an atmosphere which is light on evaluation and judgment. What causes us to work hard in such situations? Our inherent interest in and fascination with the task. This is consistent with the folklore associated with the study of creativity. What does this say to us? It says that if we can and do rearrange our schedule to spend more time on the things in which we are extremely interested, assuming that we have the time and support to keep the wolves away from our door, we will increase our creativity. It says that groups and organizations should be more creative if the involved individuals are allowed to do the same. That we should also take special effort to match people to tasks which they will be motivated to do through interest rather than through external reward.

What is wrong with relying entirely on this approach to creativity? One shortcoming is that it is counter to some traditional values in organizations. We think in terms of assigning people to work that needs to be done, not necessarily work which best matches their interests. The protestant ethic also suggests to us that perhaps work should be of a nature that requires external reward. As my father used to say, "If work was fun, somebody would do it for free." Another shortcoming is that even given activities that are so pleasurable that motivation is intrinsic, most of us do not have doors that are entirely clear of wolves. The people I know, for better or worse, are old friends with external motivation in life. I myself am a good example. In general I consider myself extremely fortunate in that I am involved in activities which bring great satisfaction to me. However, the activities which bring me pleasure are more complex than collage-making. I have not yet found a way to escape the short-term drudgery and trauma which accompany my long-term satisfaction. I am therefore continually fighting against my schedule and the clock to finish activities which are not as much fun to me as alternates I can think of (my hobbies, reading trashy novels, daydreaming). I am often aware of my income and the opinions of others. I am affected by rewards and often evaluated. Am I weak? Should I tell the world to bug off, find a rich patron, and settle down to the things I most love to do in the short term? I don't think so. I am afraid that I would lose in the long run. I am afraid that I am normal. The things which give me long-term satisfaction require short-term agony. My values seem to vote against rich patrons. I live in a world of heavy extrinsic motivation. In such a world, rewards are effective and evaluation, inadequate resources, and peer opinion are part of life.

The messages concerning intrinsic motivation are important and we should keep them in mind when dealing with groups and organizations. The more that activities result in intrinsic motivation (a feeling that tasks are done for interest and that reward and judgement are secondary) the better. To the extent that new knowledge is necessary to increase creativity (often the case) the learning should be interesting in its own right. If we need to use unfamiliar problem-solving styles, we should employ them in a way that makes experience exciting. If we need to augment motivation through reward, we should be subtle. Reward is necessary in practical situations involving normal people. Most of us are not about to increase the risk and uncertainty in our lives unless there is something in it for us. However, the reward should make us feel good about our creativity rather than simply being pay-for-labor. Successful founders of companies enjoy their wealth. However, I believe that they enjoy it as a symbol of their creative ability as much as for what it can buy. Recognition (financial or psychological) for our cleverness is always welcome and reinforces our motivation to be creative. Reward for a good chase, even though we may not make the kill, is unexpected and delightful. However, reward for accomplishment is expected and not as effective in changing our values.

Finally, as I have implied repeatedly in this book, we must be particularly careful of evaluation and judgment. They are not only liable to destroy fragile concepts, but seem to cool the fires of creativity. This is of particular importance in groups and organizations, where judgment and evaluation are not only essential, but almost a way of life.

I also believe strongly in the benefits of widespread knowledge of the problem-solving process in groups and organizations. To me the last column of the previous chart is evidence for this. Teresa Amabile explains the unusual creativity of those in this column (who were given the specific creativity "instructions") as follows:

> "For two reasons, this high creativity of the specific creativity instructions group must be interpreted cautiously. On a practical level, it is unlikely that creativity in everyday performance could be enhanced by telling people exactly what constitutes a creative performance. The reason we value creative work so highly is that we cannot know beforehand just how to achieve a novel and appropriate response. On a theoretical level, the

> conceptual definition of creativity clearly disallows the consideration of the specific instructions task as 'creative.' According to that definition, the task must be heuristic (no right answer or known technique to obtain the answer) in order for the product of task engagement to be considered creative. In this study, specific instructions on how to make a collage that would be judged as 'creative' rendered the task algorithmic. Thus, according to the conceptual definition, it is simply inappropriate to assign the label 'creative' to the performance of the specific instructions group."

I agree with her interpretation in the case of highly original concepts in individual work. But how about more pragmatic situations where many people are involved and the results must be sold to the world, the typical state of groups and organizations? In such situations it seems reasonable to me that people with a more specific understanding of creativity and its characteristics would produce outputs which would be judged to be more creative. Most of us are not Leonardo, Mozart, or Einstein and perform better in a game if we know the rules. Most of us live in a world heavy with external motivation and would be happy to make a collage which is judged creative by experts, even if given specific information on creativity. In most groups and organizations, external rewards are critical and knowledge of creativity is of great value.

Large Organizations: Origins of Bureaucracy

Keeping in mind what we have said about conceptual blocks in small-group situations, let us now look at larger organizations. Groups above a certain size, if they are to operate efficiently, need some type of formal structure. It would be difficult, for instance, for an informal gathering of 10,000 people to solve a problem quickly and economically. The character of this formal structure plays an important role in encouraging or inhibiting conceptual thinking within an organization. A convenient way to think about these characteristics is to look historically at organizational theory over the past 100 years.

With the Industrial Revolution and inceasingly complex government, organizations grew rapidly in size and in number during the nineteenth century. Toward the end of the century, theoreticians began to study large organizations and formulate principles for increased productivity. One of the first was Max Weber, the German sociologist, who in his work *The Theory of Social and Economic Organization* outlined the key characteristics of what we now call bureaucracy. Weber considered the bureaucracy the ultimate structure for large, complex activities. As he saw it, the bureaucracy operated according to a beautiful set of laws, in which the individual was subordinate to the organization, and in which structure and discipline allowed a very high degree of productivity. We are now properly somewhat cynical about bureaucracies. However, in Weber's day, the imposition of bureaucratic principles was seen as a giant step toward order and efficiency.

In the early 1900s in the United States, Frederick Taylor, Luther Gulick, Lyndall Urwick, Henry Fayol, and others developed a theory that was similar in many ways to Weber's concept of bureaucracy. This theory was called scientific management. It included concepts such as hierarchical management with unity of command (each person has a single boss) and limited span of control (originally no more than five workers reporting to one boss). Scientific management was "top-down" in that jobs were described in great detail and given to the workers by the bosses. Discipline and control were central, as were standardization of job design and performance and the accompanying uniformity of behavior. Like Weber, advocates of scientific management considered an individual within an organization to be totally replaceable.

One of the more revealing quotes from this period was a comment by F. W. Taylor about a certain iron worker named Schmidt. Taylor was dedicated to designing individual jobs in such a way that output would be maximized. He managed to redesign Schmidt's job to increase his output (loading pig iron) by some 360 percent. Scientific management subscribed to the incentive system of payment, so for this increase in productivity Schmidt's salary was increased 60 percent, which pleased Schmidt greatly and provided a happy ending to the story (obviously unions were not prevalent in those days). Taylor later made the following statement about Schmidt: "One of the very first requirements for a man who is fit to handle pig iron as a regular occupation is that he shall be so stupid and so phlegmatic that he more nearly resembles an ox than any other type."

Scientific management, despite its rather unenlightened view of the worker, was widely accepted in the early part of this century. The movement to maximize production efficiency spread to many walks of life (*Cheaper by the Dozen* is the story of an attempt to raise a family in this way). Pure scientific management was highly structured. Jobs were completely described, and no one was expected to deviate from his or her job description. The ideal employee in such a system was an automaton. It was a system that drove toward routinization and repetition and attempted to minimize differences among individuals. Scientific management was a creative development in management theory. It

was not a management approach which sought or gained creativity from the workers. It was certainly a far cry from quality circles. However, with the production-oriented economy and the large pool of unskilled and unorganized labor of the early 1900s, it was an effective system.

The "Human Relations" Movement

Scientific management's view of the worker began to be challenged in the 1920s. An influential series of experiments conducted in the Hawthorne works of the Western Electric Company underscored the importance of social and personal relationships within an organization. One particularly well-publicized set of experiments investigated the effects of illumination on productivity. In these experiments lighting levels were varied and productivity was measured in a test group of workers and in a control group that worked under standard illumination. Initially, as illumination was increased, productivity varied erratically. Productivity then began to increase in the test group, a finding that was consistent with the expectations of the experimenters, but it also increased in the control group. The experimenters then began to reduce the light level in the test group. Productivity continued to rise. At the conclusion of the experiment some of the workers were maintaining record productivity with an illumination level approximating full moonlight. The experimenters were forced to conclude that something else was going on.

A second set of experiments was run to investigate further the effects of working conditions on worker performance. In this experiment a special test room was set up in which a group of workers assembled products. In order to control the experimental conditions tightly, the workers were given periodic medical checkups and full records were taken on their sleep and other personal activities. Finally a full-time observer was put in the room to keep accurate records of all that happened. The workers were also given interviews, during which they were served ice cream and cake. Over the two years of experiments, working conditions were varied widely. However, productivity followed a general upward trend.

The results of these experiments are not surprising in this day of enlightenment. However, at the time the obvious conclusions were revolutionary: apparently people produce more if they feel noticed (the Hawthorne effect) and if they are warmly and humanely managed (the observer in the second experiment became the effective supervisor).

As a result of such findings, in the 1930s the "human relations" movement was launched. Management was still hierarchical, but considerable effort was devoted to raising the morale of individual workers. Attention was given to the physical environment and to the nature of interaction between bosses and workers, as well as to other factors that would make the worker feel noticed and comfortable. Companies hired psychologists, and organizations began to be viewed more as collections of individuals than as structures with replaceable human components.

The attack on scientific management by organizational theorists, sociologists, psychologists, and organized labor reached a peak in the 1940s. Since that time many forces in society have made the simplistic approach of pure scientific management unworkable. Workers will no longer accept automaton status. Technological sophistication has resulted in entire classes of workers who are transient, in the sense that their professional skills are in such demand that they can—and do—move easily from company to company. Such workers must be treated well if they are to be retained. The increasing complexity of external company ties (to government, to other companies, to the public) and the complicated nature of markets in our affluent society have caused organizational concerns to widen beyond internal efficiency.

Since the time of the human relations movement, many other theories of management have gained popularity. Much attention was paid to "management by objective," an approach whereby people were given objectives to accomplish rather than precisely defined duties. Organizational behavior experts have placed great emphasis on collaborative styles of management. The apparent success of such approaches in Japan has validated this emphasis.* Management theorists now write glowingly of the "organic," "informal," and "nonbureaucratic" organizations—reflecting a move toward decentralization of power and increased emphasis on the problem-solving ability of the individual, no matter where he or she appears on the organizational chart.

Creativity within Bureaucracy

Does this mean that our organizations are moving away from bureaucracy and formality? The answer is yes and no. We are certainly discovering that the management philosophies of the early 1900s, or even the 1950s, are not sufficient. However, large organizations still require formal structures. Scientific management and bureaucracy still exist. Even the most informal organization is usually able to produce a hierarchical organizational chart showing unity of command. Limited span of control is usually recognized, along with the necessity of describing jobs closely enough so that employees who resign or retire can be easily replaced and financial reward can be demonstrated to be equitable. Most organizations are authoritarian in nature and "top-down" in their policy-making structure. Time-and-motion studies and production quotas are still widely used.

In fact, bureaucratic approaches are quite effective in organizing a large number of people, and, as organizations age and grow, they tend to become even more bureaucratic. Although no manager wants his or her freedom of initiative reduced, it is an unusual manager who does not attempt to routinize the areas under his or her control. Growing organizations introduce economies of large scale, along with the necessary accompanying control devices. They mechanize and reduce the skill

* The Topeka plant of General Foods, the Kalmar plant of Volvo, and the Brookhaven plant of the Packard Electric Division of General Motors are examples of large-scale applications of collaborative management in manufacturing. These plants are organized by work group and give large amounts of management responsibility to the hourly workers, who are typically responsible for scheduling, day-to-day management of production, and often such functions as material inventory, hiring, and firing. They are often paid according to the number of tasks they master and are given the option to rotate jobs. These plants tend to show high morale, low absenteeism, and high product quality.

level of their labor force in order to increase productivity; they establish ever-increasing formal communication channels (monthly reports, weekly meetings, and so on). Individuals within the organization, including those in top management who are aware of this process, often look back fondly at "the good old days" and attempt to devise techniques to retain some of the flexibility, excitement, and motivation of the smaller organization. However, the bureaucratic trend is relentless.

The increasing inhibition of creativity can also be relentless if strong measures are not taken. In particular, problems result from increasing structural formality, control, conservatism, depersonalization, and inertia. A few comments are in order concerning each.

Structural formality inhibits communication and results in fiefdoms which can become protective of their turf. Free communication and the ability to combine activities in new ways are necessary for creativity. Perceived problems must be questioned, knowledge must be transmitted, and new and fragile concepts must first be brought to a state of reality and then sold to a conservative organization. Structural formality is usually accompanied by increased authoritarianism. In authoritative systems individuals attempt to perform well according to their job descriptions. But how many job descriptions contain the phrase "take risks"? Structural formality is also associated with routinizing, decreasing uncertainty, and increasing predictability. These may be healthy directions for business as usual, but not for creativity.

Large organizations necessarily devote a major amount of energy to control in order to be able to deal with the uncertainties inherent in complexity. It is not too difficult in any large organization to find people whose job is to prevent mistakes. Preventing mistakes involves reducing risk, which is also at odds with creativity. Because of this necessary control, the type of manager it attracts, and the more global responsibility of large organizations, they are conservative. An organization such as General Motors has far more difficulty betting itself on a new product, service, or direction than a small start-up. Large organizations are by their very nature resistant to new products, services, and directions.

They can also be depersonalizing, and here one runs directly into the motivational problem in creativity. Large numbers of people may be operating under extrinsic motivation and rewards therefore become critical. Individuals and groups must be recognized for creative and innovative output. However, often in large organizations standardization of reward system dominates and the individual prospective innovator is lost. It is not uncommon for successful founders of companies to bemoan what they see as a loss of creativity in their now-large enterprises. It is instructive for them to compare the financial and psychological reward systems in effect during the start-up phases of their companies with those in effect during the mature phases.

Finally, large organizations can be too slow. The layers of procedure and control often are not consistent with the unpredictability of creativity. Resources are needed to allow creative developments. This is especially true in large organizations because concepts must be taken to a stage of development which will allow them to combat the conservatism endemic in the decision-making structure. They must be made available when and where they are needed. Lack of time, money, people, and facilities to pursue new concepts can paralyze creativity in any large organization.

If large organizations are so fraught with characteristics which tend to inhibit creativity, what can be done to increase it? In my opinion, part of the practical answer is to ensure that particular groups of people be freed from harmful bureaucratic constraints, rather than to attempt to modify the entire structure. Creativity in large organizations is simply in conflict with many other necessary functions and must be considered in that light. Research and development groups, marketing groups, advanced-planning groups, task-forces, and feasibility study and preliminary design groups are good examples of situations where groups of people are to some extent shielded. Usually such groups are not only managed much more collaboratively and informally, but often are given control over their own immediate support needs and occasionally even placed in separate locations. In these units, power is less centralized, professionals and specialists have more to say, communication is informal, and supporting resources are more immediately available without authorization from above.

Such groups must contain the necessary links with the remainder of the organization so that their outputs will become reality. There are too many examples of research and development groups that lose their influence over the product line. They are seen as a budget drain by operations people, who criticize them as not being in the "real world." They retaliate by dropping their opinion of the intelligence of the operations people, who they term as being "neanderthals." There are innumerable examples of plans that have not been followed and feasibility studies that have not been implemented. The problem is often one of linkages. It is essential in any "creative" group that those who must implement and sell the output be represented, through complete faith in the group's competence, if not through actual participation. Healthy organizations move their best people between operating functions and activities such as planning and feasibility studies. Resistance to this (often by managers) can result in shelves of concepts that will never see reality.

Another part of the answer has to do with overall knowledge of the creative problem-solving process within the organization. The message contained in the first seven chapters is certainly pertinent. Groups and organizations reflect the conceptual blocks of their individual members: they have limited perceptions, are constrained by their cultures, respond emotionally to problems, are affected by their environments, and are limited in the intellectual strategies they employ. However, groups and organizations also provide a social framework and procedural constraints that can reinforce their members' conceptual blocks. As we have said, individuals may avoid the risks of conceptual thinking in order to protect their affiliation within the group and behave in ways antagonistic to creativity in order to satisfy their egos. Large organizations employ bureaucratic practices that are both useful for control and inhibiting to creativity. In such organizations, people may find the inhibiting effects of formalized communication and support, strict hierarchical and authoritative management, and an unresponsive reward system so stifling that they turn elsewhere to satisfy their creative urges.

Still, groups can provide a rich store of information, perceptions, technical expertise, motivation, and emotional and cultural resources to the creative individual. Organizations can accomplish things that individuals and small groups cannot. In order for this to happen, those involved in the organization must be aware of and control those factors which inhibit creativity. Formal techniques can help, and management theory now reflects a profound awareness of these problems and offers numerous suggestions as to how to handle them. However, the topic of management of creativity in groups and organizations is a major one and cannot be handled in detail in one chapter. I am in fact in the process of writing an entire book on the topic.

To me, without question one of the most important elements in managing creativity is each individual's keen awareness of common blocks to the conceptual process: their causes, their effects, and ways to overcome and sidestep them. May this book help you in that regard.

READER'S GUIDE

As stated previously, there is no complete and scientifically verified explanation of thinking which can result in universal rules for conceptualizing more productively, nor is there likely to be until a much more complete understanding of the mind is available. Still, a great deal of interesting reading exists on this subject with a sizable number of hypotheses, ranging from simplistic to elegant, which shed more light on the creative act. The reading also contains a good bit of conjecture and many value judgments which the reader must sort through and accept or discard depending on whether he is convinced or not (and probably on whether or not the reader's own opinions and values are reinforced).

Be forewarned that my opinion in this area is anything but "the final word." I have often gotten myself into trouble by assigning readings to a class and telling them to skip certain portions because they are of less value. Invariably one (or several) of the students will read these portions (presumably to see *why* they are of less value) and tell me that they are the most important parts of the reading. Therefore take my comments with a grain of salt and spend your effort reading what seems of most value to you.

I am going to give you the names of books, since they are generally more available than papers. Some are in practically all libraries and on the shelves of your local bookstore. Others are more difficult to find. Most of them contain bibliographies of books, papers, and other materials, so you will have no trouble working my few suggestions into an extensive list of future readings.

General Overviews of Creativity and Thinking

A relatively large amount of writing on creativity occurred between 1955 and 1965. As an indication of this, in 1964 Taher A. Razik, then with the Creative Education Foundation at the State University of New York at Buffalo (directed by Sidney Parnes), published a bibliography of creative studies which contained 4,176 items, half of which had been written since 1950. There were several national workshops on creative thinking in this period and an established peer group of educators working on the subject. This activity was perhaps in part catalyzed by unhappiness over the conventionality of the 1950s, which gave rise to such critics as Vance Packard, Paul Goodman, and William Whyte—who bemoaned the lack of individuality and inventiveness in society. Many of the people involved in the study of creativity during this period are still active, but the intensity of the work seems to have fallen off somewhat, due to changes in national interest, funding patterns, and the apparent resurgence of individuality (the "do your own thing" philosophy). A good look at the work that was going on in the 1950s and early 1960s is contained in *A Source Book for Creative Thinking*, edited by Sidney J. Parnes and Harold F. Harding (New York: Charles Scribner's Sons, 1962). This book contains a collection of 29 articles and addresses and 75 research summaries. Contributors include Abraham Maslow, Carl Rogers, Alex Osborne, Sidney Parnes, John Arnold, Frank Barron, and other luminaries of creativity study. The book also contains a discussion of bibliographies on creativity available at the time of publication.

Another general book for this perspective is *The Art and Science of Creativity* by George F. Kneller (New York: Holt, Rinehart, and Winston, 1965). This is a small and easily read book. It is a good introduction to the subject of creativity, the people in the field, and their discoveries. The first four chapters discuss many existing theories on creativity. The final two chapters contain the author's opinions concerning education and the nature and nurture of creativity.

One of the most widely quoted books on creativity is *The Act of Creation* by Arthur Koestler (New York: Dell, 1967). Kneller calls *The Act of Creation* "the most ambitious attempt yet made to integrate the findings of a range of disciplines into a single theory of creativity. In this book, bold in its ideas and profusely documented, Koestler seeks to synthesize his own theory of the nature of creativity, as manifested in humor, art, and science, with the latest conclusions of psychology, physiology, neurology, genetics, and a number of other sciences." The book is indeed ambitious *and* successful. It is not light reading. Over 250 articles and books are referenced in the text and contained in the bibliography (an excellent one). It is the type of book which will be read from cover to cover by those deeply interested in conceptualization. Those only mildly interested in conceptualization may never quite finish it, but always think that they should. However, you will benefit from even a partial reading.

A book which contains some excellent articles by major figures in the study of creativity is *Creativity and Its Cultivation*, edited by Harold H. Anderson (New York: Harper & Row, 1959). This volume contains addresses presented at various symposia on creativity held at Michigan State University in 1957 and 1958. Contributors include Erich Fromm, J. P. Guilford, Ernest Hilgard, Rollo May, Margaret Mead, Carl Rogers, Abraham Maslow, and others.

A book which accomplishes a similar goal is *Perspectives in Creativity*, by Irving A. Taylor and J. W. Getzels (Chicago: Aldine Publishing Co., 1975) which presents contributions from significant people in the study of creativity who were involved in a symposium on the Future Implications of Creativity Research at the Center for Creative Leadership in 1973.

Two books which are also often quoted in literature on creativity have to do with qualitative studies of creative people. *The Creative Process*, edited by Brewster Ghiselin (New York: Mentor, 1963) is a collection of writings by extremely creative people (Mozart, Einstein, Poincaré, etc.). These letters speak to the creative process used by the writers. They cover many phenomena (such as Mozart's ability to hear music and then simply write it down) that are folklore among those who treat creativity in educational establishments. *An Essay on the Psychology of Invention in the Mathematical Field* by Jacques Hadamard (New York: Dover, 1954) is a study of creativity among outstanding mathematicians and scientists. It is an attempt by a mathematician to explain mathematical and scientific invention. Two books which contain a mixture of comments from people who study creativity and writings by famous creative people are *The Creative Encounter*, by Rosemary Holsinger, Camille Jordan, and Leon Levenson (Glenview, Illinois: Scott, Foresman and Company, 1971), and *The Creativity Question*, by Albert Rothenberg and Carl R. Hausman (Durham, North Carolina: Duke University Press, 1976). Both of these books present a wide variety of viewpoints and are filled with interesting and provocative writings and opinions.

Psychological Theory

Much creativity theory is directly drawn from psychology. Therefore there is a large amount of general psychological literature of interest here. If you have never taken an introductory psychology course (Psychology I) you should read a good general introductory book to psychology in order to learn the words, the concepts, the names, and the fundamental theories. Two good ones are *Introduction to Psychology*, by Ernest R. Hilgard, Rita L. Atkinson, and Richard C. Atkinson (New York: Harcourt Brace Jovanovich Inc., 1979), and *Psychology*, by Henry Gleitman (New York: W. W. Norton and Co., 1981).

An excellent book which was mentioned in the text of this book is *The Social Psychology of Creativity*, by Teresa M. Amabile (New York: Springer-Verlag, 1983). This book is not only a study of motivation in creativity, but does a marvelously thorough job of referencing pertinent psychological studies. It is technical in its approach, but straightforward to read.

A book which is easily accessible to the lay reader is *the Psychology of Consciousness*, by Robert E. Ornstein (San Francisco: W. H. Freeman and Company, 1972). The author discusses psychology from a "left-brain, right-brain" viewpoint and makes an argument for the integration of conventional psychology, physiology, and what he calls "The Traditional Esoteric Psychologies" (psychologies which treat the "left-handed" aspects of thinking and subjects such as meditation and physiological self-control). Written as an informal textbook, it is well-referenced and well-organized. The bibliography is an excellent reading list for those wanting to learn more about the nature of various states of consciousness.

Two of the more familiar names concerned with the study of creativity are J. P. Guilford and Frank Barron. Guilford studied thinking from many perspectives and did especially interesting work on the correlation of intelligence and creativity. Samples of his thinking can be seen in *Way Beyond the IQ* (Buffalo, New York: The Creative Education Foundation, 1977). In this book he presents many measures of intelligence and discusses their relation to creativity. A classic book by Barron is *Creative Person and Creative Process* (New York: Holt, Rinehart and Winston, Inc., 1969). Those seeking a short and readable summary of psychological theory that is pertinent to creativity should try to find a copy of *The Psychology of Creativity*, by Margaret Gilchrist (Melbourne: Melbourne University Press, 1972).

If you are a fan of the writing of Erich Fromm, read *The Forgotten Language* (New York: Holt, Rinehart and Winston, 1962). Like most of Fromm's books, it is written with a minimum of referencing, footnoting, and bibliography. However, it is concerned with the unconscious and is a good discussion of symbolic language and dream interpretation. It also compares the theories of Freud and Jung with regard to dream interpretation. If you would like to compare the theories of Freud and Jung yourself, I suggest *On Creativity and the Unconscious* by Sigmund Freud (New York: Harper & Row, 1958) and *Man and His Symbols*, edited by Carl Jung (New York: Doubleday, 1964). The first is a collection of writings by Freud that are concerned with cultural and humanist matters. Many of the selections in the book deal with the particular problems of creative people and reflections upon psychoanalysis and are therefore only marginally pertinent to conceptualization. However, part of the book does treat conceptualization specifically and the book is an interesting insight into some of the lesser-known interests of Freud. *Man and His Symbols* is a translation of Jungian psychology into language accessible to the lay reader. It was put together by Jung and several collaborators during the last years of Jung's life and contains excellent presentations of the principles underlying Jung's thinking and discussions of the nature and importance of man's symbols. It is enjoyable reading. Those of you interested in a broad selection of Freud's writings might enjoy *The Basic Writings of Sigmund Freud*, edited by A. A. Brill (New York: Random House, 1938).

A specific treatment of creativity is found in *Neurotic Distortion of the Creative Process* by Lawrence S. Kubie (New York: Farrar, Straus and Giroux, 1966). Kubie explains the nature of neuroses and creative thought and the interaction between them by use of the concepts of the unconscious, preconscious, and conscious minds. His approach results in a readable and convincing argument that creativity correlates with mental health. A slightly different argument is made by Robert A. Prensky in *Creativity and Psycho-pathology* (New York: Praeger Publishers, 1980). In this book he leaves room for the well-known highly creative not-so-normals.

An excellent treatment of memory and related functions written from the point of view of the cognitive psychologist is *Human Memory*, by Roberta L. Klatzky (San Francisco: W. H. Freeman and Company, 1975). This is a textbook covering the nature and functioning of memory. It is, of course, heavily referenced, and although it is not easy reading, it is well organized and well written.

It is worth mentioning an organizational psychology book, since creativity in organizations is an overwhelmingly important topic. I suggest *Managerial Psychology*, 4th ed., by Harold Leavitt (Chicago: University of Chicago Press, 1978) which is a classic book dealing with psychological aspects of management situations. The only criticism I have ever heard of this book is that it is so well written that important concepts sometimes seem so straightforward as to appear trivial. It is an outstanding book and deals with individuals, small groups, and large organizations. Although its emphasis is on behavior and on the psychological well-being of the individual, the extension of the material in the book to the subject of thinking is simple.

Finally, read *Please Understand Me*, by David Keirsey and Marilyn Bates (Buffalo: Prometheus Nemesis, 1978). This is a book based on the Myers-Briggs Type Inventory Test, which measures problem-solving preference according to the theory of Carl Jung. The Myers-Briggs test is one which I use often in dealing with mental habits. It is non-threatening, revealing, and valuable in working with creativity in groups, since a group can become more creative merely by adding or more efficiently using divergent problem-solving styles.

The Mind/Brain

The mind/brain is the machinery of creativity, and the more knowledge we have of it, the better. A large number of interesting books which explore the workings of mind/brain are now available. One is *The Amazing Brain*, by Robert Ornstein and Richard F. Thomas (Boston: Houghton Mifflin Company, 1984). This is an easily readable and nicely-illustrated book which at the least leaves the reader feeling more aware of the machinery within. Another book which causes the reader to confront the impressive abilities and curious weaknesses of the mind/brain is *The Universe Within*, by Morton Hunt (New York: Simon and Schuster, 1982). The author is a professional writer who managed to cram an impressive amount of information about current brain research into a single volume.

Finally, a particularly thought-provoking book on the mind is *The Dragons of Eden* by Carl Sagan (New York: Random House, 1977). This book is aptly subtitled *Speculations on the Evolution of Human Intelligence*. It is testimony to Sagan's own mind, in that he collects material from an impressively wide range of disciplines and sources and combines them beautifully in order to look at the mind in the context of its past development, present role, and possible future.

"How To" Books

There is a plethora of books promulgating self-therapy approaches to psychology. In recent years, there have been major attempts to find ways of understanding and changing behavior that are more direct and less time-consuming than traditional psychoanalysis. These approaches have all featured a theory (usually not a very scientific one) of behavior and methods or exercises intended to let one understand and influence his own behavior. I find these books interesting, fun to read and experiment with, and in fact useful in understanding and influencing (in a minor way) my behavior. Do not expect to pick up a paperback at the airport, read it, and thereby shed your neuroses. However, many of these books will at least cause you to think.

There are a similarly large number of "How To" books on creativity. Many of these have been written by people who were not in the field of psychology. A reasonable explanation of this is found in *The Experimental Psychology of Original Thinking* by Wilbert Ray, who speaks of a "psychology vacuum." He says, "Political scientists postulate something called a power vacuum, which seems to mean that if the most powerful nation in a particular geographical area loses its power, some outside nation will come in and take over control. The same sort of thing happens in psychology. If there is a topic which the psychologists will not, or do not, discuss, someone else will discuss it. . . . This seems to have happened with regard to original behavior."

Two of the more well-known early techniques to enhance conceptualization are brainstorming, devised by Alex Osborn, and synectics, devised by Synectics Inc. *Applied Imagination* by Alex Osborn (New York: Charles Scribner's Sons, 1953) discusses not only brainstorming, but also Osborn's thinking about creativity. It is written in an expansive, early 1950s style and tends to reference opinion rather than science, but it strikes a responsive chord in some readers (I am not one of them). It was quite influential, as was Osborn himself, in the upsurge in the study of creativity in the 1950s. *Synectics* by William J. J. Gordon (New York: Harper & Row, 1961) talks about the early history of the Synectics Research Group in Cambridge, Mass., the technique of synectics, and its application. It is a mixture of technique and of philosophy. The book is particularly interesting because of its emphasis on metaphorical thinking.

Those of you interested in reading more about the conscious application of intellectual strategies should enjoy *How to Solve Problems* by Wayne Wickelgren (San Francisco: W. H. Freeman and Company, 1974). This is a book that presents a number of approaches to formal mathematics problems. Each approach is described and well illustrated with examples and applications. The book is well organized and written and represents an excellent example of the point of view that strategies should be carefully selected before one plunges into a problem.

An outstanding "how to" book is *Experiences in Visual Thinking* by Robert McKim (Monterey: Brooks/Cole, 1972). This book not only goes deeply into imagery in a very effective and experiential way, but also contains generous discussion of conceptualization, drawing, and general problem-solving. It is an excellently designed book with plenty of illustrations, puzzles, experiments, and problems. It is used as a text in an extremely popular undergraduate course at Stanford (Viz-Think in the local jargon) and is a convincing argument for the importance of visual thinking. It is fun to read and deceptively profound.

Although it is not really a "how to" book, *Higher Creativity*, by Willis Harman and Howard Rheingold (Los Angeles: Jeremy Tarcher Inc., 1984) is an interesting discussion of creativity. Harman is the president of the Institute of Noetic Sciences, Howard Rheingold is a human behavior columnist, and their book includes conjectures upon more esoteric motivations than one usually finds in creativity books.

There is a large and growing number of books combining exercises with "rules" of behavior. I will mention six of them. There are more. They all have different styles and are aimed toward different audiences. They are worthwhile in that they are both stimulating and enlightening. Their only shortcoming is in their tendency to emphasize the "how" to the exclusion of the "why." They are *The Universal Traveler*, by Don Koberg and Jim Bagnall (Los Altos, California: William Kaufmann Inc., 1980), *Wake Up Your Creative Genius*, by Kurt Hanks and Jay A. Parry (Los Altos, California: William Kaufmann Inc., 1983), *A Whack on the Side of the Head*, by Roger von Oech (New York: Warner Books, 1983), *Use Both Sides of Your Brain*, by Tony Buzan (New York: E. P. Dutton, 1976), *Training Your Creative Mind*, by Arthur B. VanGundy (Englewood Cliffs, New Jersey: Prentice-Hall, Inc., 1982), and *Imagineering*, by Michael LeBoeuf (New York: McGraw-Hill Book Company, 1980). I read all of these books as they come off the presses. They are generally fun and interesting and constantly remind us of the commonly accepted tactics of improving ideas.

Miscellaneous

There are many other books which cause us to think about thinking and therefore to conjecture upon the nature of the creative process. Two fascinating ones by Douglas Hofstadter are *The Mind's I*, done in collaboration with Daniel C. Dennett (New York: Basic Books, 1981), and *Metamagical Themas* (New York: Basic Books, 1985). The first is an investigation of self, and very pertinent to those who think about thinking. One cannot go very far in this pastime without coming across the elusiveness of the concept "I." The second is based upon the column which Hofstadter wrote for *Scientific American*. It is full of brain-stretchers, puzzles, and enigmas.

Speaking of puzzles, if you like such things look in the mathematical section of your libraries. Mathematicians have a soft spot for such pastimes. There is even a *Journal of Recreational Mathematics*, put out by the Baywood Publishing Co. Most puzzle and game books cater to a wide range of mathematical prowess. A few of the better ones are: *Mathematical Teasers* by Julio A. Mira (New York: Barnes & Noble Inc., 1970); *Mathematical Games* by C. Lukács and E. Tarján, translated by John Dobai (New York: Walker and Co., 1968); *The Magic of Numbers* by Sydney H. Lamb (New York: Arc Books Inc., 1967); *The Unexpected Hanging and Other Mathematical Diversions* by Martin Gardner, editor of the Mathematical Games Department of *Scientific American* magazine (New York: Simon and Schuster, 1969); *536 Puzzles and Curious Problems* by Henry Ernest Dudeney, edited by Martin Gardner (New York: Charles Scribner's Sons, 1967); and *Martin Gardner's Sixth Book of Mathematical Games*, from *Scientific American* (San Francisco: W. H. Freeman and Co., 1971). Even if you are uncomfortable around mathematics, you should still find material in such books that will keep you off the streets for a while.

INDEX